The Honey
HANDBOOK

The Honey HANDBOOK

A GUIDE TO CREATING, HARVESTING AND COOKING WITH NATURAL HONEYS

 KIM FLOTTUM

APPLE

First published in the UK in 2009 by
Apple Press
7 Greenland Street
London NW1 0ND
United Kingdom
www.apple-press.com

ISBN: 978-1-84543-302-4

10 9 8 7 6 5 4 3 2 1

Design: Yee Design
Cover Images: (main image) Courtesy of the National Honey Board; all other photographs
by Kim Flottum with the exception of front, bottom middle, www.istockphoto.com.
Illustrations: Robert Leanna II
Recipe Photos: Glenn Scott Photography
Food Styling: Catherine Kelty

Printed in Singapore

To Kathy –
For patience,
For help,
For love

Contents

4

Fragile:
Handle with Care

5

Using What You and the
Bees Have Made:

Varietal and Handmade
Honey Recipes

Preface

ONE OF THE BEST THINGS about being a beekeeper is harvesting what your bees and you have produced. You and your bees worked hand in hand to grow strong, forage great distances, and finally put away that most perfect gift from the bees—honey. Harvesting and using your honey is the last stage of this exciting task.

You can, like the industrial beekeepers, harvest a season's blend just once a year, and you can use harsh chemicals to chase the bees away, followed by heating and filtering your honey crop to *death*. Or you can follow a simpler path, an easier way, and finish with a product that people will never, ever forget.

The Honey Handbook is the first book of its kind, focusing on the best ways to produce, harvest, and use what you and your bees have created.

Your bees produce not just *pure honey*, but an even more delectable, more desirable, rare and select delicacy—hand-made honey, sometimes called varietal honey, but always, always premium honey.

We are going to create something new. Think of this for a moment … do vintners create and sell just *pure wine*? Do high-end coffee shops sell only *pure coffee*? Or do luxury creameries make and sell simply *pure ice cream*? Heavens no! And beekeepers don't produce just *pure honey*! Bee-keepers and their bees produce honeys with millions of flavors and colors and aromas. Each honey is as unique as the flowers it came from, as different as the geographies the bees roam and forage, and as complicated as a blend of a thousand perfumes. Every unmixed-by-man honey is different. Every hive produces its own secret blend. With your help, honey's day is here.

Let's see how this works.

When your bees have completed finding, gathering, curing, and storing this premium crop, their job is done—but your task has just begun. Now this fragile, delicate, delicious combination of sugars, flavors, aromas, and sweet delight needs to be collected as gently (and as often) as possible, then handled with care without bruising or burning, without harming it at all.

First, the protective beeswax cappings must be removed (without heat, without damage), and the perfectly ripened contents of these honey bee–built hexagonal chambers are for the first time exposed to the cold, cruel world. The responsibility for nurturing that innocence, that natural purity from flower to feast, now lies with you and you alone.

Next, you must urge this syrupy sweetness from these beeswax chambers—still in gentle mode and only barely, barely warmed. Extracting is such a harsh term, but that's what it's called. Extracting honey from the honey comb is much like spinning water from lettuce and collecting the water; the mechanics are identical, only the equipment is larger. The intricate honey comb, now empty and barely used, is returned to the bees so they can fill it again with yet a different mix of this premium product.

When the honey leaves the confines of the extractor, it is strained (and never, ever filtered) to remove the remnants of beeswax and fragments of frames that sometimes come through.

Next, it settles quietly, undisturbed in a warm and dark place, and in a week or so the surface is cleared and cleaned of any air bubbles that lurk within. What remains is bright and clear and ready.

Some then warm this mix, but to no more than the warmth of a lover's touch, and quickly decant into bottles and jars to sell or share, while others wait and store their treasure until needed later.

While waiting to be used, this mix may relax, reducing itself to its simplest state—the crystal. All honeys do this in time—some soon after harvesting, some months or even years later. It has to do with the sweet sugar chemistry inside. Honey is sugars and water, aromas and flavours, colours and minerals, and all stay together when honey crystallizes. Some beekeepers guide their honey to that natural state intentionally, and we'll carefully examine that process. Natural crystal honey is smooth and creamy like apple butter and doesn't run or drip or pour. It is unique and too-seldom made.

And then comes the very best part—using these unique and one-of-a-kind honeys to lend their special qualities to breads and drinks, to sauces and desserts, to main meals and simple snacks, and to-die-for things you can make to eat and savour and share.

The Honey Handbook takes you through the process of moving honey from beehive to honey house, revealing and extracting it so that none of the finer aromas, tastes, or colours are bruised, burned, or broken. Then we look at careful, considerate storing, and finally, we use all the kinds of honey you can make or find in a whole cookbook-full of tasty and healthful recipes.

What, I ask, could be better?

No Longer a Beginner

THE PREMISE OF THIS BOOK is that you have attained a degree of success in raising honey bees: keeping them healthy, replacing and increasing your colonies, perhaps raising your own queens, making splits or divides successfully, and doing all these things with a minimum of expense and effort. Once you've arrived at this point—and you know whether you are there or not—it's time to refine the operation. This book looks at a fundamental of beekeeping—producing a honey crop that is not simply the accumulation of an entire season's efforts from your bees, but rather producing specific and defined crops of artisan and varietal honeys.

Remember, vintners don't just produce pure wine; they start with specific grape varieties and process these varieties with precise fermenting and aging techniques. We are aspiring to do something similar with your honey crops, no matter whether what is generated is only a few jars every season, or hundreds of jars three or four times a season. These varietal honeys will be better than anything you have ever tasted.

For many people, honey is more of a chore than a luxury. It can get in the way; it's heavy, sticky, and messy; and maybe you don't even like it or can't eat it. It may be an obstacle to managing your pollinating honey bees. One old-timer once observed that people get into beekeeping because of the bees, but get out of it because of the honey. It happens, and that's okay.

In my first book, *Complete and Easy Guide to Beekeeping*, it was suggested that honey was actually a by-product of the pollination process and was (almost) disposable. Individual frames or whole supers could be given to other beekeepers who wanted the honey. More drastically, it was suggested that those full frames of honey could actually be discarded—thrown out with the trash. That is not an option exercised often, but it's not much different from the surplus zucchini grown and discarded every summer by many avid gardeners.

MORE FEARSOME THAN A SWARM OF BEES?

It's absolutely true that without bees there's no honey. So there must be bees. Add to that the principle that it takes a lot of bees to make a lot of honey. So what does it take to not just have enough bees, but a lot of bees? Mostly, it takes experience, some luck, and being in the right place at the right time.

A strong, healthy colony is what is needed to make honey, lots of honey. But the beekeeper needs to be constantly aware of a colony's condition so it doesn't get so crowded, congested, and ambitious that it swarms. From the colony's perspective, being able to throw a swarm is a good thing—that means it is healthy and vigorous enough to reproduce, thus continuing its genetic line, the goal of all living things.

A swarm can also be a bad thing: the colony has reduced its workforce by half and probably won't be as productive during the season had it not swarmed. The swarm itself, however, will have incredible energy and will produce comb and brood and honey the first year if the weather and good health permit.

A swarm of bees is a sign of a healthy, vigorous colony as well as a harbinger of lowered productivity.

WHY KEEP BEES?

Besides honey, there are numerous reasons to keep bees: pollination of the garden, a fascination with nature, a family project, or an interesting hobby that pays for itself. A honey bee colony is a new world to explore, learn from, and enjoy, whether it is fancy or plain, expensive or average.

But beekeepers who expand their operations (from two colonies to four, ten, or fifty) eventually invest in harvesting equipment that includes fume boards or bee blowers, uncapping knives or machines, tangential or radial extractors, storage pails or tanks, and bottling equipment. This is the natural progression of things, not unlike a gardener whose gardening operations expand to include more space, mechanized tilling equipment, and automatic irrigation systems.

Beginning beekeepers, those who typically start their adventures by purchasing a package of honey bees, tend to spend their first bee season wrestling with the basics of feeding, supering, maintaining honey bee health, and preparing for winter. Often their fledgling colonies do not produce and store enough honey for overwintering, and end-of-season feeding is a necessary task. This is the fault of neither the beekeeper nor the honey bees, and it is an essential learning experience for the new beekeeper.

A backyard apiary may be neat and pristine...

❧

...or it may become a tad weedy over the summer...

❧

...or appear to be abandoned, so far back in the yard.

❧

And then there is the beeyard that appears somewhat disheveled, but the bees are happy, healthy, and thriving.

❧

But almost as often these brand-new colonies produce a honey surplus—more than the bees need to overwinter—and the new beekeeper must quickly learn to harvest the surplus. For a rookie, this first honey experience can be exhilarating, whether it's the first, second, or even the third year into this activity, or it can be a disaster, depending on what the new beekeeper has had a chance to learn.

FINALLY, THE NEXT STEP

Squeezed cut-comb, chunk, and comb honey introduces beginning beekeepers to the world of honey production without their having to invest in expensive uncapping, extracting, and bottling equipment, and many beekeepers never leave this mode of production. They enjoy these products for themselves, share them with receptive family and friends, or find a market for them.

There are many techniques for coaxing honey from the honey comb. Some are quite destructive to the hive's comb and are costly to the bees, but they are common techniques for beginners because they require a minimum investment in equipment and time.

Squeeze and Drip for Liquid Honey

The simplest honey-harvesting technique is the squeeze and drip technique. It consists of removing honey-filled combs from a hive (this process is discussed on pages 64 to 66), cutting the beeswax comb from the wooden frame, and then placing the chunks of comb in a fine-mesh bag. The comb and bag are squeezed, and the honey is collected in a pail. Aside from honey, this method leaves crushed beeswax and no honey comb in the hive—effective but messy, time-consuming, and wasteful. The beeswax comb has been destroyed and will have to be replaced by the bees. (Note: This technique will not work for beekeepers who began their adventure with a starter kit with plastic foundation in the frames.)

Cut-Comb Honey

Beekeepers can harvest entire frames with the comb and honey still intact, and cut the honey and comb from the frame into smaller pieces. The excess honey can be drained from the cut edges. These pieces, now called cut-comb honey, are honey (and wax) at its purest, almost untouched by human hands, and left just as the bees made it. It is exactly what the bees eat, and has all of the flavours, aromas, vitamins, minerals, pollen, and color that it was meant to have.

Chunk Honey for Liquid and Comb Honey

Pieces of cut-comb honey can be placed in a jar, and the remaining space in the jar can be filled with liquid honey. Called chunk honey, it is highly prized by honey connoisseurs. Some beekeepers choose liquid honey extracted from the same nectar source as the honey in the combs to pour into the jar, so the honeys are identical. Others mix two kinds of honey; a stronger, darker variety in the comb surrounded by a lighter, milder variety creates a unique taste experience. Besides, the jar's presentation is very attractive.

Round Comb Honey

Using somewhat more sophisticated equipment, any beekeeper can produce comb honey using Ross Rounds, which are round plastic sections available from most bee supply dealers. This is also a destructive production technique, but not at all messy. A white ring fits snugly in the prepared holes in each half of the brown frame. Comb honey foundation is placed between the two halves, and they snap together. The bees build comb on the foundation, filling the holes in the frames. When the honey is capped, the frames are removed, and the two rings (now securely fastened) create one round section. A cap is placed on each side, the round is labeled, and the unit is ready for sale.

HIVE HINTS

Choose the Right Foundation

If you are going to produce cut-comb, chunk, or comb honey, be sure to buy the beeswax foundation made especially for these products. It is very thin—thin enough to read a newspaper through. It's made exclusively from wax cappings, and no wires are installed in the wax (nor will you add any when inserting the foundation into the frame). Install this foundation as close as possible to the time the frames will be added to a colony; otherwise, the thin foundation will droop, causing bee space problems. Use foundation pins to hold the foundation in place. Once the bees have built up the cells, the resulting honey comb built on that foundation will stay in place.

Heated Beeswax

Beeswax won't be harmed at 140°F (60°C), even if it is held there for some time. Heating to a very hot temperature, more than 175°F (80°C), or holding beeswax at liquid stage for hours will darken the wax and drive off some, if not all, of the aromas for which beeswax is known.

GETTING THERE

That first honey crop, oh, how sweet it is. No matter how the first honey crop is gathered, it will be the most flavorful and aromatic honey that friends, family, and neighbors have ever tasted. And it doesn't matter whether bee parts, beeswax, frame fragments, or air bubbles are in this honey—the first time you glide your finger across the surface and lick that warm, golden nectar, you'll taste it, smell it, feel it, even hear it ... it's truly a sensuous experience. Every beekeeper worth his or her hive tool remembers this first time.

The second year, and probably the following two or three years, beekeepers refine their technology and hive management techniques. They figure out the tools, the timing, and the discipline of keeping bees healthy and productive. Because of this initial attention to harvesting details, the finished product—the honey itself—often takes a backseat simply because the logistics of collecting, extracting, and bottling capture all the attention. Each year there's a lot to learn, and the window for experience is short.

During the first few years, there are other distractions, too. Colonies die for no apparent reason. They swarm and don't produce any crop at all. There's lousy weather—too much rain, not enough rain, too hot, too cool. These factors all compete for attention and time and weigh down the learning curve of becoming a beekeeper.

The first year or two is a stressful time for a beginning beekeeper. The dropout rate during the first years is often high. The original expectations were, perhaps, a tad overzealous, or the family's commitment wanes because children grow, space is commandeered for other uses, or the day job changes. Simply, life happens.

This weeding process can be difficult and expensive, but ultimately it's Darwinian. Once that process has run its course, those that remain—the true beekeepers—have mastered the basics and are finally equipped for honey to take center stage.

To prepare cut-comb honey, first remove the comb from its frame, and then cut the comb into pieces sized to fit into containers. (Cutting comb with plastic templates will consistently reproduce the correct size comb.) The cut comb can be left on a screen or mesh wire board overnight, draining the cut edges of honey, before packaging.

☞

Preparing chunk honey is very similar to preparing cut-comb honey. Cut the comb using plastic templates so the cut piece or pieces fit neatly in the jar. Place the cut comb in the jar, then pour a light-colored liquid honey, such as alfalfa or sweet clover, to fill the jar.

☞

Many beekeepers use beeswax to make items for the home and family, such as beeswax soap, hand lotions and creams, and other cosmetics.

☞

Round comb honey in Ross Rounds—plastic comb honey frames, rings, and covers is another production technique.

☞

Chapter 1

Where and Why Bees Forage

THE SIMPLE ANSWER to where the bees are, of course, is in the backyard. A question with a less simple response might be, "Where do bees go when they leave the hive to forage for nectar, pollen, water, and propolis?" The myriad answers to this question are the background for much of this book.

Experienced beekeepers generally know the production cycle of their backyard colonies. Seasonal plant progressions and landmarks are relatively standard year to year, and historical information—such as the first and last frosts—is readily available on local calendars, seed packages, and almanacs. From these two general events some management decisions can usually be made—add supers right at last frost, harvest a month before first frost, and the like.

CHANGING LANDSCAPES

Foraging destinations do not remain static throughout the season. Seemingly idyllic landscapes are constantly churning with growth, destruction, rebirth, and other changes from the first hint of spring to the last breath of autumn. Farm crops bloom, then turn to fruit, or the field may be replanted and bloom again. Fields that were productive last season may be fallow this season, inviting a variety of dormant weed seeds to sprout and bloom. Meanwhile, some areas are bee-barren in the spring but burst into bloom in midsummer or even fall. Still other areas are loaded with bloom in the spring but revert to grasses for the rest of the year. Herbicides are routinely applied to farm fields, roadsides, and public areas, knocking down current and potential blooms; yards are mowed; woodlots are removed or trimmed; or a new development named Willow Creek (where neither willows nor creeks spring forth) suddenly appears. The list goes on and on.

As a result of all this change, bees will not forage in the same place for very long. If they do stay in a particular location, it is because the flowers change; they will be foraging on different flowers when the first crop matures and disappears.

ALFALFA HARVEST

Alfalfa (Medicago sativa) *is one example of an unpredictable, here-today, gone-tomorrow, back-again-in-a-month honey plant. Farmers generally cut alfalfa for hay when about 10 percent of the plants are in bloom because it is protein rich and makes the most nutritious hay for their cattle. If it rains for several days just before a planned cutting and the farmers can't proceed, the field will soon be full of alfalfa blossoms. The bees will win this round; they should harvest a large crop of excellent honey.*

To see just how good the nectar is, a recruit will ask for and receive a taste of the nectar the returning forager carries. If impressed with the quality of the nectar and the vigor of the dancing forager, the recruit may choose to go to the advertised flower patch herself and bring back another load.

While one area close to a colony may be profitable—a high-quality food reward that is only a short distance from the colony—it may be limited in how many bees it can support at any one time, or how much pollen it will produce over the duration of bloom. This is known as the carrying capacity of a flower patch, and every group of flowers has a maximum capacity. Carrying capacity is influenced by how many honey bees or other nectar and pollen collectors are visiting the patch. Honey bees may be visiting from other colonies—either yours, bees from another beekeeper, or honey bees from feral colonies—or other nectar and pollen foragers such as bumblebees and solitary bees may be competing for the rewards, too. Depending on the resident and visiting insect population, only so many foragers from your colony can take advantage of this particular crop.

Bees from the same colony may simultaneously visit one, two, three, or more patches, even though they may be closer or farther from the colony, or the patch may be more or less profitable. These may be different areas with the same crop blooming (think dandelions *[Taraxicum officinale]* in the backyard or nearby park), or different crops blooming in the same location (think willows *[Salix* spp.*]* and yellow rocket *[Cruciferae* family*]* on a woodlot hillside). These permutations present an ongoing and often continuous smorgasbord for your bees, but the menus at each of the restaurants are slightly different.

FOOD COLLECTION AND STORAGE

The dynamic of food-gathering varies by location, time of day, day, and season all at once. Each colony is doing its best to track these changes using information relayed back to the colony by regular foragers returning with information on current crop conditions, and by scouting foragers who are exploring as much of their environment as time and bee-power allow.

The driving force of this dynamic is not a commanding general, a board of directors, or even a queen. Further, there are no experienced, retired foragers from last year to show everybody how the system works. What makes all of this happen, every year with every colony, is simple chemistry.

Driven by Hunger

When the queen begins laying eggs in the spring, the brood she produces sends a constant chemical message to the nurse and house bees in the colony that there are hungry mouths to feed. The more brood there is, the more brood pheromone they send out, thus the stronger the message. Initially there's usually (but not always) ample stored pollen and honey in the hive, so this demand can be met. When those stores begin to run low, and they usually run critically low later in the spring when brood rearing really ramps up, nurse bees, house bees, and returning foragers react to this "feed-me" message quickly and with purpose.

Fed by Pheromones

Although "feed-me" is a major force to reckon with, another message often exists in a colony: the presence of empty honey comb. The natural hoarding instinct of standard European honey bees dictates that if there is empty comb in the colony above the brood nest

Honey bees remain in the "feed-me" larval stage for about five and a half days, but in that short time they outgrow and shed their cuticle five times, increase in size and weight nearly 600 percent, and are fed many thousands of times by the nurse bees in the colony. They don't create excrement until the very end of their larval stage, just before they stand upright in their cells and spin their cocoons before pupating. During their "feed-me" phase, they produce a pheromone that tells the nurse bees that yes, this is a larva and you should feed this larva. This "feed-me" message is what drives the colony to collect food, to store food, and to protect itself from enemies.

☞

Different stages of larval growth require different diets. What the house bees feed these larvae varies every day, as there are all stages of larvae in a colony all the time.

☞

Nurse bees eat pollen for its protein content and honey for carbohydrates. These nutrients, plus additional enzymes and other chemicals from their feeding glands, are combined into a mix that feeds worker larvae the first three days. (Incidentally, it is similar to royal jelly—literally, the diet of queens.) After the first three days, they are fed a mix of honey and pollen called bee bread. It has less protein content and less sugar, which acts as a feeding stimulant. Workers eat a reduced-quality food, and eat less of it—one early distinction made between workers and queens.

☞

After only four days, the larvae fill two-thirds of the cell they were born in. Tomorrow they will stand upright, void the complete contents of their guts, quit eating, and begin to spin their cocoons.

☞

Flower Fidelity

Bumblebees and other bees are opportunists and will visit any flower close to their nests.

When floral sources are mixed in the same location, individual honey bees will visit only one source during a foraging trip and ignore the other. This behavior is called flower fidelity, and this makes honey bees the best pollinating insects around.

Honey bees practice flower fidelity with nearly religious fervor. When a forager arrives at a flower patch, such as an apple orchard (*Malus* spp.), and she has either visited the location previously or has been recruited by a scout who tempted our forager with the sweet smell of apple blossoms, the forager does not hesitate to commit. She will fly from one apple blossom to another to another, ignoring the dandelions' nectar and pollen. This is beneficial on many levels. The apple trees (and the apple tree grower) are pleased because the forager is transferring apple pollen from one tree to the next, which is critical for each apple blossom to set fruit. If a bee went from apple to dandelion to a clover (*Trifolium* spp.) patch across the road, nectar would be collected but no flowers would be pollinated.

Further, the nectar she collects during her monofloral trip is all apple nectar, which, when mixed with the apple nectar brought back to the hive by other foragers, will produce a concentration of varietal honey strictly from the orchard. Other bees in the same colony may be visiting the same location, but dining exclusively on the dandelions, ignoring the apples completely, aiding the dandelions in their quest for immortality.

Another benefit of this behavior is that each pollen variety brought back to the hive is different in its composition of proteins, enzymes, lipids, fats, vitamins, and minerals. If a colony had only apple pollen to dine on, a nutritional deficiency might develop. Bees, like humans, need to consume a balanced diet.

Synthetic Brood Pheromone

The prototype of the brood pheromone dispenser

Empty comb in the colony is a stimulant almost as strong as the message sent by brood. It simply screams, "Collect more honey!"

The newest management trick honey bee scientists have concocted is to produce a synthetic brood pheromone that can be placed in a colony. It sends the same message as real brood pheromone: "feed me." This signals the colony to forage more than it would given its amount of brood.

Even if there is no brood, the colony begins foraging. Foraging loads are larger for both pollen and nectar, more foragers seek both pollen and nectar, and house bees begin foraging at a younger age to help fill the demand created by the addition of this synthetic chemical. The result is increased foraging for both pollen and nectar, and more fruit set because of increased visitation to blossoms. An added benefit is that the population of the colony increases because more food is available to feed more brood.

It seems like a win-win, but field trials are still ongoing to measure this benefit to the beekeeper, the grower, and the bees themselves.

It is recommended that synthetic brood pheromone be used in the spring to increase colony population for pollination jobs, again in the summer to increase forager population to collect more nectar for honey, and once again in the fall to guarantee a large population of bees to go into winter. That said, constant population enhancement will absolutely lead to swarming, so that, too, must be routinely monitored.

area, it must be filled with honey, and as much honey as possible, as soon as possible. Honey is survival food for the winter, and without stored honey the colony will die. Bees won't stop gathering honey as long as there are plants to produce it, bees to collect it, room to cure it, and a place to store it for later. If any one of these is missing, the system breaks down and honey production falters until fixed, or fails for the season if not fixed.

Successful foragers returning with a nectar load offer their bounty to several recruit bees to taste so they have an idea of both the odor and the value of the load. House bees then unload the forager and take the nectar to the honey storage area of the colony. The greater the demand for food the faster returning bees are greeted and unloaded.

If demand slows for any number of reasons, foragers quit foraging. If there's no place to cure or to store the nectar, no brood to eat the honey or pollen, or not enough house bees to unload the returning foragers of their nectar loads, returning foragers get this message pretty quickly. No use wasting time and energy on an unnecessary task.

Those foragers who are returning with pollen must unload their own bounty, but there's always some remaining on their bodies that house bees can sample. Pollen, like nectar, varies in quality, quantity, odour, and colour. When the pollen find is large and the quality good, the message to the bees inside is strong and vigorous. If the find is subpar, the enthusiasm expressed by the returning forager will be weak, too. But more important, these same foragers are strongly influenced by the demand created by the brood in the hive—that demanding brood pheromone is a mighty motivator.

Three small clusters of bees each communicate with a returning forager. When foragers or scouts return to the hive with a nectar load, they head to an area of the hive commonly called the "dance floor." Here they encounter house bees that will relieve them of their load and take it to dehydrate, plus they find foragers and recruits looking for foraging tips. If the demand of empty comb is strong, the returning forager will be unloaded quickly and there will be many recruits seeking directions. Each individual bee will decide to follow, or not to follow, the instructions given based on the need of the colony, the value of the reward, and the vigor of the returning bee.

☞

Often a back door to a colony is created by a crack at the top intended to improve ventilation or as the result of ill-fitting equipment. The more entrances to a colony, the more energy a colony spends defending its entrances. Bees of guard-age could be foragers if the colony didn't need them watching the door. Also, the more entrances, the more opportunity for colony robbing, for foreign bees that are carrying pests or diseases to enter, and for predators to gain access.

☞

The pollen clinging to this forager (center) gives the recruits a sample of the quality, the taste, and the smell of the flower it came from—all a recruit needs to find the flower patch.

☞

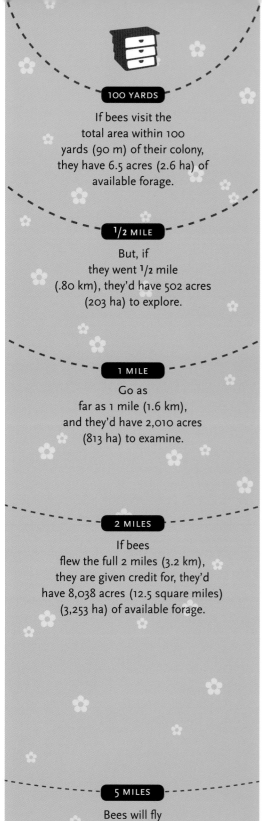

100 YARDS

If bees visit the total area within 100 yards (90 m) of their colony, they have 6.5 acres (2.6 ha) of available forage.

1/2 MILE

But, if they went 1/2 mile (.80 km), they'd have 502 acres (203 ha) to explore.

1 MILE

Go as far as 1 mile (1.6 km), and they'd have 2,010 acres (813 ha) to examine.

2 MILES

If bees flew the full 2 miles (3.2 km), they are given credit for, they'd have 8,038 acres (12.5 square miles) (3,253 ha) of available forage.

5 MILES

Bees will fly farther if there's no forage nearby. If they went 5 miles (8 km), they'd have 50,240 acres (79 square miles) (20,331 ha) to gather from.

Where the Bees Go

The amount of ground that bees cover depends on how far from the hive they must fly to find food. The hive's location may inhibit flight in certain directions because of non-foraging conditions such as lakes, industrial areas, or monoculture farm crops that don't produce honey or pollen (wheat, oats, or rice). Alternatively, they may have complete freedom to go wherever they want.

Honey bee foragers will travel only as far as they must go to obtain the nectar, pollen, water, or propolis they seek. Scout bees are an exception because by definition—and necessity—their scouting missions tend to take them into uncharted territory, where they seek their nectar, pollen, water, or propolis.

Once their treasure is found, scouts gather their booty and return home. At the hive they explain to their nest mates where their food samples are located using odour, taste, and distance and directional information. If the colony has multiple foraging choices available because many scouts have been successful in finding food, recruits will follow those that appear to have found the most rewarding locations. Recruits are many in the summertime, when rewards are rich and the living is easy.

When a scout honey bee has located a rich but distant food source, or a less productive (in terms of amount or quality) but very close source, her urgings may be somewhat lethargic, and recruits will be less willing to follow her if more enthusiastic scouts are dancing at the same time. The cost of visiting any of these sites is expensive—the formula of energy expended versus energy gained applies.

Conversely, if there are very few or no nearby food sources, then distance and reward become less important because need within the colony takes precedence. Bees need to eat no matter what, so every food source is a good source.

Insight into Hive Activity

In an average-size colony in late spring, without extremes in bad weather or problems within the colony, and in a location providing adequate to good foraging, what's going on?

First, there is a lot of activity at the front door of the colony. Scouts are returning, loaded with nectar and pollen messages for the foragers. Regular foragers are returning with treasure troves of pollen and nectar collected from flower patches, to which they had been sent earlier by an energetic scout or other foragers. At the same time, empty scouts and foragers are heading in the opposite direction, leaving the colony in their quest for more food.

Traffic Flow Inside and Out

If there is a lot of food available in the wide, wide world, the traffic at the door can be impressive, even for a moderate-size colony. While food gatherers are coming and going, the guards inspect as many bees as possible so only those bees with security clearance gain entrance. Other foragers (those collecting water or propolis) are also coming and going, while undertaker bees and trash removers are in and out concentrating on their tasks. Meanwhile, young bees are taking their first practice, or play, flights.

Inside the colony, traffic is no different, but the environment is of a completely different orientation from the outside world. Hive activity is vertical, unlike our horizontal world. Work surfaces are up and down, left and right. Also, it is absolutely pitch black inside—no light penetrates the interior of a colony. Holes and cracks are covered with comb or propolis to keep both invaders and light at bay.

Once a returning forager moves more than a short distance away from the front door (or any other entrance), the temperature and humidity can rise dramatically when the hive is in a temperate environment. The coolest parts of the hive are about 85°F (29°C), and in the brood nest it rises to higher than 90°F (33°C). Humidity is close to 100 percent, and bees are walking on, over, and next to one another—a multidimensional subway at rush hour.

There are aromas by the hundreds, perhaps thousands, that each bee takes in, analyzes and acts on, reacts to or ignores: the smells of incoming fresh pollen and nectar, yesterday's nectar being reduced to honey, and cured honey and pollen already stored. Each scent is unique in itself, and the maturation process alters the expression of these aromas over time.

Add to this a multitude of pheromones being emitted from the queen. Most significantly, during the rush of summer in a healthy colony the brood pheromone dominates, affecting every bee in the colony. Usually the banana-scented alarm pheromone from guard bees at the entrance or at an opening is in the air. There's always some Nasonov pheromone inside and outside all of the time—all this headiness in the hot, dark, crowded, busy world of a hungry, bustling honey bee colony.

Traffic at the front entrance can be pretty congested during a honey flow, and guards can be hard-pressed to challenge everybody that wants in. Foragers from other colonies with a food load that have drifted to the wrong colony gain easy entrance because of their payload. Drones drift, too, and gain easy access during a flow. Only when the flow has subsided do the guards begin to get serious about who is invited in.

Air Flow

Ventilation is important during a honey flow. Foragers return with nectar that is roughly 80 percent water and 20 percent sugars and need to reduce the percentage to about 20 percent water and 80 percent sugars. Fanning at entrances and inside brings in cooler, lower-humidity air from outside, which replaces the warm, higher-humidity air inside—as the air warms, it takes up water and the nectar is rapidly reduced. It generally takes six to twenty-four hours to reduce a day's haul from nectar to honey if there's adequate ventilation and enough bees to move all that air.

Chapter 2

Guide to Honey Plants

EXPERIENCED BEEKEEPERS KNOW there are several management styles that can be used to reach any particular set of goals for the season, but generally these goals all have some things in common.

SPRINGTIME MANAGEMENT FOR HEALTHY BEES

First, above all, you want to keep your bees healthy all of the time. There are so many things you can do with a healthy colony of bees, but there's almost nothing you can do with a colony that's not up to par. And to be strong during the active part of the season, the colony has to be strong during the less active times—after the honey flow in the fall and all during the winter season. Except in the most tropical environments, honey bees have a downtime when the outside world does not provide a living, but even then, they must remain healthy so they are ready to gear up and get going when the next active time arrives.

Maintain Enough Bees

Second, you want the population of your colony to be just right at just the right time, and you want that to happen with as little stress on the colony as possible. If your colony naturally grows when it needs to, the probability that they will be able to produce an optimum honey crop is very high. To accomplish this means that you need to be keenly aware of when the major honey flows occur where your bees are, and provide the care necessary to prepare the population so it peaks, or at least is very strong, before most nectar is ready to harvest.

There are at least three schools of thought regarding the concept of boosting the population in your colony, and conscientious beekeepers should examine all of these before beginning a management regimen that you (or your bees) may regret at a later time.

**TAKE CARE
OF THE BEES**

With beekeeping, you are always looking at least two seasons ahead. Here's a foolproof piece of advice: take care of the bees that take care of the bees that go into winter. It takes healthy bees to raise healthy bees, and there is no compromise in that.

All is well in the world when spring trees finally bloom and the weather cooperates, but getting a colony ready to take advantage of this feast can present problems for you and for the bees.

THE "LET-ALONE" SYSTEM. The first to consider is essentially a let-alone system. This doesn't mean passively managing the troubles that bees can encounter; rather, it means letting the colony's population expand at the rate governed by the weather, available stored food, and their own genetics. Italian honey bees tend to go full speed in brood rearing, no matter the weather or available food outside. To sustain this rate of growth, there must be enough stored food or they will surely starve before outside food sources become available. Carniolans and Russians, however, are careful about their reserves, maintaining a slow and careful brood-rearing pace until outside food is abundant. Whether the colony reaches an optimum population in time to take full advantage of the nectar crops available is second to the fact that they have dovetailed their schedule into the schedule of Mother Nature. This is certainly the least work for the beekeeper, though it may not be the least stressful situation for the bees because the bees may actually starve before they can get enough food to continue.

THE "BEEKEEPER AS NATURAL FOOD PROVIDER" SYSTEM. A second management scenario is similar to a let-alone system; however, the beekeeper plays a more active role in ensuring all is well in the colony nutrition-wise. Rather than let

Protein Supplements

If you plan to feed protein to your bees, the best choice is pollen that was trapped the previous season. Store the pollen in the freezer until it is needed, then blend 3 parts pollen to 1 part honey, until firm. Some beekeepers will mix pollen, pollen substitute, and honey or sugar syrup at a 1:1:1 ratio. Place this food on the top bars, and do not let the supply run out.

the bees completely fend for themselves, with their growth and expansion regulated by available resources, a beekeeper will intervene by keeping enough of the right kind of food in the colony *all of the time*. And what kind of food is that? The best food there is: frames with honey and pollen stored from the previous season given to the bees continuously, *before* they need it, so they don't run short, so brood doesn't starve, and certainly so the colony as a whole doesn't starve. The key is that it is *only* pollen collected last season and *only* honey made by your bees. (The food in frames is best, but pollen that has been collected in a trap and mixed with a liquid honey base makes a nutritious and attractive free lunch.) These are the foods that honey bees have eaten for millennia, and they are given only *before* the colony needs them. This way, from the bees' perspective, there is never a break in the food supply.

If this management plan is undertaken, you are making an intense commitment. When you become the food provider—that is, Mother Nature's cupboard has gone bare and there's nothing left inside or available yet outside—you must never, ever quit. When a colony's natural rhythm becomes dependent on you, you have supplemented Mother Nature. And as long as the colony determines there is a certain level of food available, they will undertake spring growth and expansion plans at a rate supported by that amount of food and their normal growth pattern. They commit by producing brood at an increasing rate due to lengthening photoperiod and warmer weather while relying on that steady food source. It is as if they had, indeed, stored more than enough food last fall.

But if suddenly the food stops (did you plan that vacation just now on purpose, or what?), they can, for a time, sacrifice themselves to feed brood by converting the protein in their own bodies to food for their young. When prolonged, this cannibalism can be incredibly stressful on the adult bee population. That said, when food hasn't been incredibly abundant, most colonies won't overdo brood production (even the prolific Italians), keeping relatively even with the food supply. A beekeeper who is keenly invested in this commitment, and reliably provides, will be rewarded with a potent, robust colony that remains unstressed and healthy. This is the goal that is at the heart of being a beekeeper, and this management style is encouraged above all else.

THE "ARTIFICIAL CALENDAR" SYSTEM. There is a more intense management style that is not uncommon, but to some it is unhealthy and certainly stressful for both the bees and the beekeeper. This technique involves beekeepers imposing a completely artificial calendar on their bees. They begin feeding their bees long before the colony is ready to produce young as determined by photoperiod (lengthening days) or outside temperatures. The food of choice is pollen substitutes and supplements available from beekeeping suppliers. This food is made from soybeans and milk by-products, minerals and vitamins and whatnot, all ground to a fine powder. High-fructose corn syrup is added, and the concoction is formed into a patty, which is placed in the hive on the top bars. How bees find this mess and why they eat it is bewildering, but it stimulates them to begin feeding the queen in earnest so she begins laying in earnest and the colony is off and running, preparing for another season—way, way out of sync with the outside world—and they become as breathless as you are after reading about this.

That this food is not honey bee food by any stretch and that the calendar they are now living by is completely askew (relative to photoperiod and outside temperature) creates stress in the hive. In addition, like the second method, the colony is absolutely dependent on the beekeeper to maintain and gradually increase the amount of food placed in the colony at each feeding, and there are shorter time periods between these feedings. As the population increases, the amount of food consumed increases—until finally Mother Nature steps in and the outside world becomes a friendly place again. This is an intense, unhealthy, and unnatural technique for raising bees. It can produce a sizeable colony just when the beekeeper demands it, and if honey production is the absolute goal, this is one way to achieve it. Perhaps, though, it's not the best way for the bees.

SUMMERTIME MANAGEMENT FOR HONEY PRODUCTION

There are many, many manuals, articles, and books that explain numerous ways to successfully manage a honey bee colony during the summer season in most parts of the world. We've already looked at some spring management decisions you can make that use—or abuse—your bees. Moreover, these techniques can be tweaked to accommodate a beekeeper with a few colonies in the backyard, a few dozen in beeyards around the county, or a commercial operation with hundreds or even thousands of colonies spread out all over (see Resources, page 164). However, let's examine in detail a couple of techniques that highlight summer management situations you may encounter during a typical season.

But let's first define a *typical* season. For starters, a typical season is predictable. There aren't any weather extremes—three months without rain or only three days it didn't rain; or a temperature regimen twenty degrees below normal for the first half of the growing season or twenty degrees warmer than usual all season long.

Second, a typical season doesn't have any major honey bee health surprises or queen emergencies. A healthy colony begins the season with a small, overwintered population and rapidly grows during the early spring season in preparation for later nectar flows. The early nectar crops then provide more of the food necessary for increased brood production and colony growth (without swarming). This growth, if left to its own devices, too often peaks *after* the main nectar flow in most regions. That's because the bees build their population *on* the flow, instead of

The greatest threat to a successful season is an invasion of varroa mites. Ongoing non-chemical treatments, such as sugar dusting and drone trapping, coupled with always using bee stock that is mite resistant, and the occasional treatment with soft chemicals, such as essential oils, should keep mite populations in check.

Monitor populations continuously during the season, however, using sugar shake, sticky boards, or alcohol wash to make sure your bees stay safe.

Dividing strong, populous colonies in mid- to late season is becoming a routine practice. Brood production is interrupted, reducing varroa populations; splits can be requeened with fresh, new queens to lead colonies into winter; and the mother colony can be requeened, left as is, or joined to other, small colonies for the winter. This is an ideal time to apply soft varroa treatments (if needed) to the original colony, since no more honey will be harvested from that colony that season.

before the flow. It's typical for colony growth but less than ideal for the beekeeper wishing for a good honey crop. Moreover, it certainly leads to swarming behavior if not checked early in the cycle. It is why many beekeepers artificially stimulate their colonies early in the season to meet the goal of peak population before bloom.

This type of growth curve levels off during the summer, however, and eventually decreases again to an overwintering level. That final population will depend on the race of bees being used, with the darker bees typically having much smaller populations and little to no brood than the prolific Italians that produce bees and brood year-round. This typical growth scenario is probably the most common management program beekeepers use. It requires the least manipulation, management planning, and work, and it produces, most years, an adequate honey crop. But it isn't the only one.

Another Plan: Divided Colony

Another practice is becoming more popular that both serves as a mite management program (this is the "keep your bees healthy" part) and assists in queen production for beekeepers in northern areas. Here, a colony begins the season as above, but during the doldrums of summer, this now-large colony is divided. It may be divided into several smaller colonies, depending on the population of the original mother colony. A new queen (local or purchased; mated or queen cell) is added to each divided colony (and any needed pest control strategies are applied), which then grows rapidly and heads into winter with a strong, young population. At the end of the season, the mother colony has produced several healthy crops of honey, including, perhaps, a late summer and fall crop, and goes into winter as the old lady in the apiary—and the future is riding on the splits, not the mother colonies. Next spring, surviving mother colonies are used again if healthy, or joined to the new, strong, and healthy splits to produce a large, ready-to-roll colony.

This technique has the advantages of aiding in mite control by breaking the brood cycle (denying varroa a place to reproduce); introducing a productive queen into a strong, young colony; and providing an opportunity to check pest problems as well as prepare for next year when the weather is favorable, food is optimal, and other management chores are most easily accomplished. Even the mother colonies are aided by this because they have already made what honey crop is available and still have both young bees and older bees to gather a fall crop and to go into winter well stocked. If you haven't considered this management technique, it may be something to explore. It's becoming more common, especially in moderate and northern climes, and it has shown itself to be beneficial in a number of ways, especially when producing varietal honey crops.

If a healthy colony unexpectedly experiences a varroa mite explosion, becomes overrun with small hive beetles, or suffers any of the other maladies that can, and sometimes do, get out of hand, the rules and the management plan must change to save the colony. That usually means sacrificing at least part of the honey crop so treatments or management schemes can be applied. This the price of maintaining healthy colonies without using harsh chemicals, and is not unlike staying home from work for a few days when you have the flu.

WINTERISE WELL

Finally, a beekeeper wants to provide everything a colony needs to successfully prepare for the winter season. Preparations for next year begin even before the first crop has been harvested *this* year. You need to make certain you have controlled the typical pests in your colonies at the right time so that only healthy bees go into winter and the exotics such as varroa and their viruses are nonexistent. Make certain that you leave your bees with more than enough of the right kinds of food—honey and pollen, not sugar and soybean meal—and that they are protected from the elements, including the wind, rain, and snow, plus pests such as mice and other small critters.

MANAGEMENT SUMMARY

Regardless of the management techniques used, the goal is to have as many forager-age bees as naturally possible in your colony that are ready to go when nectar is available. Usually that's early in the season, but a midseason or even a late season flow may be all that you have. So the first step is to know when the nectar flows. Sounds simple, right? Let's see.

The Best Sources of Information

The best source of information about when things bloom where you and your bees are tends to be anyone who makes his or her living on that nectar flow—your local, experienced beekeeper. By necessity, these beekeepers have good information about what blooms when, and most are experienced enough to make a good guess from year to year. For instance, in the region in which I live, I can say with confidence that the first flush of dandelion bloom—that is, those early blooms on the south sides of buildings in protected locations—will be the last week in April. Knowing that, with equal confidence I can say that the first day of what I call field bloom will be within the first ten days of May. Thankfully, there are a couple of other ways to make that same prediction, other than through the accumulation of decades of experience.

First, keep good records. When I first moved to my current geographic region, I kept good, consistent records for several years. And every year, dandelions had

Plan ahead for a safe winter for your bees. Make sure there's plenty of food and protection from the elements, and that the colonies are free of pests and problems. This windbreak of trees is great insurance for keeping these hives healthy.

field bloom sometime during the first ten days of May. After a few years my record keeping flagged—I had internalized the calendar, or so I thought.

But extended warm or cold spells and other weather events cause the dates of first bloom to change. Warmer weather may produce earlier bloom, and an extended cold spell or a late snow can delay bloom by days, maybe many weeks.

Does a week or two make much of a difference? Well, when bloom lasts only a week or ten days, you've missed getting ready by half of the bloom if it's early. And that's half of the honey crop you were planning on or your bees needed for buildup.

Good records serve a beekeeper well, but only up to a point. The calendar method of bloom prediction has the advantage in that it's easy, but the disadvantage of being only moderately accurate. Most beekeepers, however, can live with that uncertainty. Or they just assume that the bloom will be early and plan accordingly. That way they don't miss it.

One confusing aspect of the calendar technique is predicting when one nectar source is finishing and another is just beginning. If the first crop started late, or the next crop comes in early, your bees will be gathering both at the same time. If you are producing a blended spring blossom honey, this works to your advantage because anything that comes in at about the same time tends to be alike—most spring blossom honeys tend to be light and mild. However, if the goal is to separate the dandelion bloom from the later (and richer) black locust bloom, your efforts may fail. (Mixing dandelion and locust honeys, incidentally, borders on the criminal.)

Accurately predicting bloom dates is an exact science that crop growers have been honing for years. Apple, blueberry, and almond growers, and producers who need to know when their crop will bloom, must know exactly when those flowers will show up *before* they show up. A growing degree day program has been specifically designed to accommodate the crops in question. Crop organisations, university and government agencies, and crop consultants routinely track growing degree days for farmers so that they are exactly right on for crop bloom.

Honey producers don't have these resources provided to them specifically for honey crops, but existing data can be used by beekeepers to their advantage, or you can do your own growing degree day record keeping. See the technique in the sidebar if you don't already know this trick.

To calculate bloom date using growing degree days, first do a little research. There is an incredible amount of information available on the Web to growers of corn, apples, and other crops. Simply search for "growing degree days" and your state or region and you will find an abundance of data.

CALCULATING GROWING DEGREE DAYS

....

There's a simple technique for starting your own growing degree days (GDD) database, but first a ground rule. Plants essentially don't grow if the ambient high temperature for the day is more than 86°F (30°C), or the low temperature for the day is below 50°F (10°C), so these two extremes must be noted in all calculations.

The basics of calculating GDD is as such:

Using various resources or media (such as weather websites or networks), find the high temperature for the day, add it to the low temperature for the day, and calculate the average by dividing by two. Subtract that number from the base temperature you are using (which is 50°F [10°C]).

$$GDD = \frac{Max.\ Temp + Min.\ Temp.}{2} - Base\ Temp$$

If the high temperature for the day was 84°F (29°C) and the low was 60°F (16°C), the calculation would be:

$$GDD = \frac{84 + 60}{2} - 50$$

$$72 - 50 = 22\ Growing\ Degree\ Days$$

Remember the extremes. If the high for the day was 90°F (32°C), the low 40°F (4°C), the calculations go like this:

$$\frac{86\ (replacing\ the\ 90) + 50\ (replacing\ the\ 40)}{2} - 50 = 65 - 50 = 15\ GDD$$

The first question should always be, "When do I begin calculating this figure each season?" The simple answer is: a few days before the last frost date in your area. This annual event is available from gardeners, Extension offices, most university Web pages, and governmental information sources that feature agricultural information.

Alternatively, any local program (check local university or grower organisation Web pages for this information) can provide the information, then refer to your records. The correlations are easy: When did maples bloom? When the local growing degree days reached a certain point. And they'll reach bloom when the growing degree days reach that point every year, rain or shine, sun, snow, or sleet, no matter what the calendar says.

Even though information for, say, apple growers, is being tapped, the data can be used to predict all manner of honey plant bloom dates. If you have a good history of those dates and access to past dates for agricultural growing degree days, the rest is easy. If not, a year or two of careful record keeping will provide the basics.

Multitudes of Honey Plants

A friend once cataloged the plants visited by his bees over the course of an entire growing season in the small space between his back door and the door to his honey house, only 45 yards (41 m) away. There were, he found, more than 250 different varieties of plants that his bees visited between the last frost in the spring and the first frost in the late fall.

When calculating general bloom dates in most of the Northern Hemisphere a good rule of thumb is that dandelions, which are globally abundant, tend to bloom during the first week of May near the 45th parallel. (That's Chicago in the United States, central France, and northern Italy in Europe.) Adjustments must be made for altitude, where mountaintops will be cooler and plants bloom later than in the valleys below, and for the effects of warm ocean currents, such as those that affect parts of the U.K., which will speed things along. Other local and microclimate effects can hasten or retard bloom dates from what the calendar says, which reinforces the need to establish growing degree days as bloom indicators rather than "firm" rules.

SOME OF THE BETTER HONEY PLANTS TO KNOW

So, what honey crops are you looking for? Seasoned beekeepers generally know the major nectar crops their bees make honey from, but a surprising number of beekeepers don't know just how many good crops are out there. Too often, honey shows up and they don't know the source. These mystery sources fall into the following categories:

- Tree crops (such as black locust or citrus)

- Field crops (such as soybeans or alfalfa)

- Major land-use crops (such as sweet clovers on government reserve land in the United States, or heather in northern Europe)

- High-value annual crops (such as borage in Canada, or sunflowers in Russia)

Most beekeepers know the major wildflowers that emerge during the season, such as dandelions, willows, privets, Russian olives, and the like. But there are so many different plants that bees can harvest a crop from, beginning very early in the spring and lasting to the very end of the season. Of course, in the near tropical and tropical regions, there is seldom, if ever, a dearth of blooming plants, and bees are able to work nearly all year long.

Included here are scores of plants beneficial to honey bees that provide nectar, pollen, or both. For the purposes of this book, the primary focus is plants that produce nectar, the source of honey. The lists have been grouped into several rough categories, but there are many that overlap—apples, for instance, are an agricultural crop, an ornamental in urban landscapes, and an escape from orchards and yards. These lists are provided in, very roughly, bloom sequence order, earliest in the season to latest, but it is best to determine for yourself when they bloom in your region. The same species of plant grown in different environments can be unpredictable, and only firsthand observations can confirm exact growing degree day information.

These lists are by no means all-inclusive of the plants honey bees visit. Regional and continental differences exist, and you may have all or only some of these where you keep your bees. Plants not listed may provide all or most of your crop. The core message is to find the plants, then observe your bees visiting them when they are in bloom, note the plants' abundance, verify bloom time and duration, try to identify the type of honey that is produced, and note when that honey is ripe and ready to harvest. These are the golden rules of producing varietal and handmade honey no matter where you are.

These lists provide the common name, genus and species (if only one), and the usual colour and flavour of the honey the plant or family of plants produces. Keep in mind that these descriptions are generalizations because the same species of plant can produce different types of honey in different locations due to weather, soil type and quality (which affects nutrition, moisture availability, and a host of other physiological functions), and other environmental factors. Plus, the same plant in the same location may produce a different honey year to year depending on the weather during any particular year. Dry years tend to concentrate nectar flavours, while wet years tend to mellow those same flavours.

Moreover, different species of the same genus can produce vastly different kinds of crops. A good example of this is privet (*Lingustrum* spp.). One species that is commonly used as a hedge in northern U.S. home landscapes, common privet produces a bitter, strong, and basically unpalatable honey, while its Chinese cousin that is primarily a wild shrub in the southern United States produces a mild, sweet, and very popular honey. These differences are not unlike the subtle but distinctive differences in wine made from the same grape varieties, but grown in different locales. Beekeepers can and should use these differences to distinguish their local honeys from products produced in other locations. Your place is special. Capitalize on that fact.

These lists are not plant keys. These lists are for honey type information, notes on the plants, and very general bloom dates in the United States where they grow. Countries in the Southern Hemisphere can adjust the dates by six months to account for the "reversed" seasons. Some plants do not grow in harsher northern climates, while others cannot survive in tropical areas.

Finding All Those Plants

The best way to discover where the plants that are attractive to your bees are growing is to explore on foot. Satellite maps will help discover where they might be, but being an eyewitness is the only true option. For example, golden honey plant only grows where there is moist soil, along streambeds, or on the edges of ponds. These areas tend to have dense tree cover, so the plants won't be visible on satellite maps.

IDENTIFYING PLANTS
IN THE WILD

Though photos of some of the plants listed are included, these lists are not intended to provide positive identification such as a plant key or good identification book will do. To study and identify plants in an unfamiliar environment, a good tree and wildflower reference book is essential. There are many pocket guides that are easy to use and contain loads of information on plant growth requirements, identification, blooming times, flower and fruit characteristics, and more. A half-hour in a good bookstore will solve all of your identification problems. Get as many of these as you need to aid in your hunt for the perfect honey plants where you live. You will have to rely much less on the spotty information you may receive from other beekeepers, and you will know for certain where your bees are and when to expect a honey crop from the plants they are visiting.

Identifying Honey

If you are able to identify a particular honey from a known source—black locust, for instance—place a sample in a small, sealable bottle, cover it in foil to keep out the light, and freeze it to keep it from darkening too fast. That sample won't last forever, but a photo of your fresh sample for colour check along with a sample jar of the honey itself will go a long way in helping you identify honey samples when you don't know for sure where they came from. You'll have colour *and* flavour for the record.

Tree Crops

Tree crops tend to bloom early in the season, but there are midsummer and even later varieties, too. Most of these are valuable for early spring buildup and occasionally to produce a harvestable honey crop. Because of their early blooming nature, tree crops tend to have unpredictable nectar production, affected not only by this season's weather but also by stresses from last season. For instance, the drought of last summer can affect flower and nectar production this season, a reminder of the importance of keeping good records.

BLACK LOCUST 📷—*Robinia pseudoacacia:* (shown above) A somewhat erratic crop that produces an exquisite light, mild honey with a long-lasting fruit-flavored aftertaste unlike any other honey. Unfortunately, its early spring bloom is too often hampered by inclement weather. May and early June.

CITRUS—*Citrus* spp. (includes oranges, grapefruit, limes, lemons, and others): Often simply labeled orange blossom honey, citrus produces a medium to dark honey. It is sweet with a citrus or extremely light aftertaste. This can be a tricky crop because of the pesticide sprays that are often necessary during bloom. Orange blossom honey was once a mainstay in the varietal honey market, but it is no longer quite as abundant. January into spring.

EUCALYPTUS—*Eucalyptus* spp. (There are so many varieties of these growing in so many places in the world that listing them all here would be prohibitive.) Most varieties provide a moderate to dark honey with a strong, somewhat medicinal, flavour. Generally blooms spring into summer.

BLACK MANGROVE—*Avicennia nitida:* Mangrove produces a light honey, often with a somewhat earthy taste. Limited to tropical wet areas.

TULIP POPLAR 📷—*Liriodendron tulipifera:* (shown at right) This tree is unique in both the flowers and honey it produces. In the United States it is common from Pennsylvania to northern Florida, west to the Mississippi River in the south, and Indiana and Michigan in the north. The honey is dark and distinctive but not strong. Shortly after harvest it has a reddish tinge, but that fades after time. The flavor is unlike any other honey. April to June.

BEE BEE TREE 📷—*Evodia daniellii:* (shown at right) Usually used as a specimen tree, this majestic tree blooms in midsummer. Honey is medium amber with a mild syrupy aftertaste when pure. Incredibly attractive to bees, the nondescript blooms are covered from early morning to late afternoon with visiting bees.

MAPLES 📷—*Acer* spp.: (shown at right) Maples of one species or another are common in most of the eastern United States, with some more northern, some southern. One is common in the west. Maples produce one of the very first major honey flows beginning in February in the south to May in the north, blooming early and serving mostly as support for brood production. An occasional honey crop can be made if the flow is heavy, the previous fall was moist, and the spring weather has cool nights and warm days. The honey is light, mild, and not distinctive.

Tulip Poplar
Liriodendron tulipfera

Bee Bee Tree
Evodia daniellii

Maples
Acer spp.

Willows
Salix spp.

Catalpa
Catalpa speciosa

Basswood
Tillia spp.

TUPELO—*Nyssa ogeche* and spp.: Few if any honey plants are as well known as the tupelo. In song and story, in movies and history, this tree and the honey it produces are legend. In the United States it grows mostly in the swampy areas of southeast Georgia and along the Apalachicola River basin in the panhandle of northwest Florida and southern Georgia. Historically, beekeepers moved their bees along the river following the bloom, putting colonies on boats or high docks to avoid the flooding. The light amber honey is mild and pleasant, and it carries a host of subtle floral flavors in its aftertaste. It is highly sought after, and if pure, is said to never granulate. Late March to early May.

WILLOWS 📷—*Salix* spp.: (shown at left) There are several willow tree species in North America and many worldwide that are favored by honey bees during the early spring bloom. Willow tree nectars tend toward light amber and indistinctive, but are seldom harvested because the bees use the honey for food. Pollen from these is important.

CATALPA 📷—*Catalpa speciosa:* (shown at left) Huge flowers routinely drip with nectar during bloom, but the nectar is quite watery. They are always a good nectar source, but they are even more attractive during dry spells because there is less water and more sugar in the nectar. The honey is light, it appears to be watery, and the flavor is nondistinctive. Catalpa trees are commonly planted along roadsides and as specimen trees in urban areas. June.

BASSWOOD 📷—*Tillia* spp.: (shown at left) These trees are common in woodlots, especially the big-leafed varieties in the northeast corner of the United States from Maine to Minnesota and south to Kentucky. Common as urban planting nearly everywhere, the little-leafed varieties are equally productive. Both are excellent nectar producers when the weather cooperates.

This light, somewhat minty-flavoured yet aromatic honey has a unique flavour that is not appreciated by everyone. The midsummer bloom is too often affected by summer's dry, windy weather, which desiccates the blossoms.

SOURWOOD—*Oxydendrum arboreum:* The famous though certainly common-looking sourwood tree produces one of the most distinctive and flavourful honeys in the United States. Located mostly in the Blue Ridge Mountains area, sourwood honey is nearly water-white, delicate, and exquisite, with a tiny hint of sour to offset the very sweet. Unfortunately, it is often mixed with other honey from plants blooming at the same time. Even more unfortunate: far, far more sourwood honey is sold than produced each season. Sourwood trees bloom in July in the North Carolina hills, and beekeepers traditionally move colonies to the cooler hills for this flow and to escape the heat of summer in the lower areas of the state.

GOLDEN RAIN TREE—*Koelreuteria paniculata:* Typical legume tree–flavoured honey—light, mild, and not terribly distinctive. Spectacular specimen plant when mature in the home landscape. July.

PERSIMMONS—*Diospyros virginiana:* This medium-size tree blooms in May through June and grows primarily in the southeast corner of the United States from West Virginia south to Florida and west to Oklahoma. The honey is exquisite, mild, light-coloured, and too often mixed with other floral sources, but when pure or nearly so, it's a delight.

CHINESE TALLOW TREE—*Triadica sibifera:* A common name of this invasive plant is popcorn tree because the blossoms look like popcorn. Common in the southern United States, the honey is light amber, good-flavoured, and ample most years. May.

TREE OF HEAVEN—*Alianthus altissima:* Common, short-lived urban escape, this invasive is very aggressive and very productive, but the honey is bitter and ill-tasting and can ruin a super full of otherwise good honey. Avoid if possible. June and July.

THE BASSWOOD TREE IN BEEKEEPING LORE

There is a beekeeping history to the basswood tree that should not be forgotten. In the early days of modern beekeeping (after L.L. Langstroth developed the movable frame hive), A. I. Root planted a basswood orchard on the outskirts of Medina, Ohio, and put his queen-rearing operation in the heart of it. He used the basswood for three things: the trees provided shade during the heat of the summer, and when the trees matured, the blossoms provided a nectar source in June and July. Finally, when the trees were of the correct age, they were thinned and the timber provided the special wood needed to make the basswood sections for section comb honey.

Peaches
Amygdalus persica

Soybean
Glycine max

Agricultural Crops

There are a large number of crops that are grown commercially for food and fiber that are beneficial to honey bees for the nectar and pollen they provide. Moreover, many of these plants are enhanced by or require honey bee visitation for pollination purposes. Occasionally farmers will rent honey bee colonies to be brought in during bloom to provide that much-needed pollination. This is certainly one way a beekeeper can add to the bottom line of the business.

The downside of these crops is that both during and after bloom, these same farmers may apply lethal pesticides to these crops to control the insect pests that damage or destroy them before harvest. Honey bees are insects, and insecticides kill insects. Beekeepers must be careful when their bees are in the vicinity of most agricultural crops. Pesticides are a way of life for farmers, and a way of death for honey bees.

APPLES—*Malus* spp.: Apples are an orchard crop, a common and beneficial urban landscape tree, and a specimen tree in home yards. Honey is light amber, somewhat distinctive, but very pleasant. May.

PEACHES ◉—*Amygdalus persica:* (shown at left) Attractive in early spring, this tree rarely produces a honey crop because it's used mostly to feed larvae rather than stored. However, if there is a surplus, be careful, because the honey can be quite bitter.

ALMONDS—*Prunus* spp.: Found on the West Coast of the United States and in Spain and China. Almond honey is bitter and not suitable for human palates. Noted here because it is the largest honey bee pollinated crop in the world. February.

SOYBEAN ◉—*Glycine max:* (shown at left) It used to be that soybeans were fickle about producing honey. Certain varieties grown in warmer areas were said to produce light amber, mild, sweet but nearly flavourless honey good for blending. But interestingly, reports of soybeans producing harvestable crops in the cooler north began appearing shortly after genetically modified herbicide-resistant varieties of soybeans were planted. This is unconfirmed, but anecdotal evidence suggests it's correct. Another new need-to-know aspect of this crop is the increasing use of pesticides to control introduced rust disease, and soybean aphid.

BRAMBLES—*Rubus* spp. (blackberries, raspberries, black raspberries): White to extra-light amber honey with exceptional mild fruity flavour. Commonly grown almost everywhere as a commercial or home garden crop, but also a prolific escapee, colonizing nearly everywhere except extremely arid areas. In the United States it is especially common in the Pacific Northwest. The prolific May bloom is beneficial for brood development, but if planned for, it can be an incredible early crop.

CLOVERS 📷—*Melilotus* spp. and *Trifolium* spp. (Alsike, white dutch, white sweet, yellow sweet, Ladino, red, arrow leaf, many others): Clover honey, (shown at right), is the workhorse of the honey industry. By far the most common honey produced in much of the world. In the United States, it is common nearly everywhere but the desert southwest. Although varieties differ somewhat, the honey is generally light to very light and mild flavoured, which means it is very indistinct. Clover honey is what almost everybody considers honey to be—what honey should taste like and what they are used to.

COTTON 📷—*Gossypium* spp.: (shown at right) Grown in many parts of the world. Honey is light to light amber, good flavoured. Integrated pest management programs—boll weevil eradication programs being instituted nearly everywhere cotton is grown—have reduced pesticide kills on this crop significantly in recent years, bringing cotton back as a major honey crop. Spring.

PLUMS 📷—*Prunus* spp.: (shown at right) Medium to light honey, with usually little distinctive taste. Cultivated plums are common, as are wild plums, which bloom early, usually in April, generally long before most surrounding trees, so they stand out. Mostly medium light to medium amber, mild with an undistinctive but pleasant flavour.

CHERRY—*Prunus* spp.: Many varieties, some in orchards as sweet and tart, some in yards as specimen plants, many escapes in woodlots. Surplus honey is uncommon because it blooms so early. April.

CANOLA—*Brassica* spp.: This mustard family crop produces a mild-flavoured honey that some find strong and flavourful, others nondescript. It granulates rapidly once extracted. Often blended with other crops to reduce granulation. May to July.

Clovers
Trifolium spp.

Cotton
Gossypium spp.

Plums
Prunus spp.

Trefoil
Lotus corniculatus

Alfalfa
Medicago sativa

HAIRY VETCH—*Vicia villosa:* Grown as a cover crop and as a hay crop. The honey is water-white and flavourful; some label it as bold. Several varieties of vetch are used for erosion control along road-sides, but they have no attraction to honey bees. Summer.

TREFOIL 📷—*Lotus corniculatus:* (shown at left) Once a common forage plant for cattle, and harvested for hay, this legume is less commonly used now, but is still somewhat common in the United States from New York west to Wisconsin, either planted or as an escape. The honey is white to extra-light amber and fairly strong with a distinctive, metallic aftertaste. Generally favoured where it is grown. Most say it tastes like strong clover honey.

ALFALFA 📷—*Medicago sativa:* (shown at left) This and clover are the workhorses of the honey industry, though alfalfa is not quite as common as a honey producer as it was a few decades ago. The honey is light and mild, and it has a distinctive tangy aftertaste that identifies this crop immediately. Pesticides and early mowing have reduced the beekeeper's willingness to place colonies near alfalfa fields, but occasionally poor weather will curtail mowing (usually done at about 10 percent bloom), and a fine, full crop can be had before the field dries to where it can be mowed.

BUCKWHEAT—*Fagopyrum esculentum:* Common buckwheat is sel-dom grown as a grain crop anymore, but it can be found where gardeners are producing it for green manure, to be plowed right after bloom. The honey of this variety is very strong, black, and pun-gent. Some beekeepers in the U.S. Midwest grow it specifically as a honey crop. There is another variety that is used for the same pur-poses, but the flowers produce a much lighter, milder honey, unlike anything associated with its dark, strong cousin.

PEPPERMINT—*Mentha piperita:* This herb is grown as a crop for its oil and as an herb in home gardens, and as an escape. Like all mints, it is highly attractive to honey bees when it blooms in mid- to late summer. The honey, when in quantity, is dark, strong, and minty flavoured. Otherwise, it adds a hint of mint to mixed wildflower crops.

Wildflowers

The number and variety of wildflowers that honey bees visit over the course of a season certainly varies by locality, but the number of blossoms is in the millions, and the number of varieties is certainly in the many thousands. The true value of honey bees is apparent with these varied sources of nectar and pollen. Because they are attuned to visiting a single species of plant during any foraging trip, honey bees accommodate that plant's pollination needs quite nicely. The clover blossom to clover blossom to clover blossom trip results in pollinated clover blossoms. If they traveled from clover to apple to dandelion, this pollination task would go unfulfilled. This flower fidelity is what keeps most wildflowers coming back year after year after year.

DANDELION 📷—*Taraxacum officinale:* (shown at right) Probably the best-known wildflower, or yard pest, depending on your point of view, in the world. Honey is yellowish, strong, and not very pleasant, but usually turned into bees rather than harvested. April through May blossom.

HENBIT MINT—*Lamium purpureum:* Early bloomer, usually April and May, in the eastern United States primarily. If nectar is stored, the honey is mild with a touch of mint. Unique is the orange pollen spot on the heads of returning bees. When there's a surplus, it may be mixed with dandelion, which can result in a not-unpleasant blend to harvest.

MUSTARDS 📷—*Brassica* spp.: (shown at right; includes wild mustard, black mustard, rockets, and others) These canola cousins produce a mild- to strong-flavored honey; some find it strong and flavourful, others unattractive. Granulates rapidly once extracted. Generally, collected early enough to be used only for buildup. If harvested, often blended with other crops to reduce granulation. April to July.

Dandelion
Taraxacum officinale

Mustards
Brassica

Purple Loosestrife
Lythrum salicaria

Spanish Needle
Bidens spp.

Fireweed
Epilopium angustifolium

Sunflower
Helianthus spp.

PURPLE LOOSESTRIFE 📷—*Lythrum salicaria:* (shown at left) This plant is an introduced species in the United States but common to parts of Europe. It became a nuisance in wetlands but was a prodigious honey producer in midsummer. Government intervention has reduced its spread. Honey is pleasant and somewhat heavy bodied, but has a greenish tint when first harvested. This fades about a month or so after extraction.

SAGES—*Salvia* spp. (white sage, black or button sage): The major honey producers are in the western United States in dry, arid areas. Honey is white for both, mild and very attractive for black sage, somewhat heavier and stronger for white. It is often a major producer in the western United States, but drought will reduce output.

SPANISH NEEDLE 📷—*Bidens* spp.: (shown at left) There are many species of plants in this group, and all are productive and attractive to honey bees. Honey is distinctive, but pleasant, tasting somewhat like a crushed flower smells. When produced in enough quantity, the honey makes an excellent handmade product. Species bloom from midsummer to frost, depending on location and variety.

FIREWEED 📷—*Epilopium angustifolium:* (shown at left) Sprouts up after forest areas are cleared for logging or after a burn. Thrives until shaded by new growth, so its life expectancy is limited in any given area. Produces extremely clear, water-white honey, with an exceptional, delicate flavour. Highly prized, but not common outside its growing area.

SUNFLOWER 📷—*Helianthus* spp.: (shown at left) Sunflower crops are grown worldwide, but pesticides are troublesome. Oilseed and edible varieties are available, and both are generally attractive. Hybrid parents for breeding, however, tend to be unattractive and very unproductive. Honey is medium amber, flavourful, and usually loaded with pollen. Wild types abound nearly everywhere in the temperate world, with similar honey types. Summer.

THYME—*Thymus serphllum:* Common in pastures, where it can compete with short grasses as it is not favoured by grazing cattle. This small plant produces a strong, minty, pleasant-tasting honey from mid- to late summer. The naturalized population is declining with the loss of dairy farms in the U.S. Northeast and Midwest.

PURPLE STARTHISTLE, SPOTTED KNAPWEED 📷—*Centurea* spp.: (shown at right) This unassuming wildflower resembles two or three closely related species, but all are similar in appearance, attraction to honey bees, and quality of the honey they produce. They have purple flowers, grow in gravelly, poor-quality soil on the dry side, and produce one of the best honeys grown in the eastern United States. They have spread considerably since introduced, and range in the United States from Maine to Georgia and west to the plains, but they're strongest in the northern regions. Producing even during dry years, the honey is full bodied, but flavourful and not strong, with excellent aroma, mouthfeel, and aftertaste. It is in the same genus as yellow starthistle, which is common in the western United States and produces a similar, perhaps even more popular honey. Midsummer bloom.

YELLOW STARTHISTLE—*Centurea* spp.: Different from its eastern cousin, this plant produces a thorny, spiked yellow flower cluster that is bothersome to cattle. An introduced wildflower, it spread widely and rapidly in California, but recently introduced pests are beginning to limit its production, along with persistent drought. The honey is similar to that produced by its purple cousin, but is even better, if that can be possible. Midsummer bloom.

COMMON MILKWEED 📷—*Asclepias syriaca:* (shown at right) This common weed is home to a variety of well-known insects, primarily as larvae food for the much-favoured Monarch butterfly. It blooms from June to August with large, round flower clusters. The honey is mild, but distinctive, tasting a little like one of the aromatic crushed flowers. The pollinia, baskets containing the pollen, sometime ensnarl and capture visiting honey bees.

Purple Starthistle
Centurea spp.

Common Milkweed
Asclepias syriaca

Smartweed
Polygonum spp.

Teasel
Diapsacus sylvestirs

SMARTWEED 📷—*Polygonum* spp.: (shown at left) There are a large number of plants in this genus that are excellent honey producers. The small, common smartweed blooms in mid- to late summer, but its towering cousins, the Japanese bamboos, bloom earlier or later depending on the particular variety. The latter can be 6 to 8 feet (about 2 to 2.5 meters) tall, and they produce a reddish to dark brown honey, often called blood honey or chocolate honey. Others are short and nearly never seen, but produce a light, mild, but nondescript honey. This is an important plant mostly in the east and midwest sections of the United States.

LING HEATHER—*Calluna vulgaris:* The premium honey plant of northern Europe is nominally established in the northeastern United States. Honey is strong, somewhat reddish, and slightly aromatic. It is the migratory beekeeping crop of Europe, where beekeepers move bees to the moors where the plants grow wild for the bloom, then return home. It has a nearly cultlike status where it is abundant, not unlike sourwood or tupelo honey in the United States.

HORSEMINT—*Monarda* spp.: Common in the central south region of the United States, this mint blooms from May to July. Once an important producer, it has been reduced in importance and population by effective herbicides, but it's still common in neglected and unused land. Honey is medium coloured, with a minty flavour that's unique when mostly pure. If possible to isolate, it makes a truly wonderful varietal honey.

JOE-PYE WEED—*Eupatorium maculatum:* Large stands are becoming more common in abandoned meadows along tree lines if the location tends toward damp. The mild amber honey is unassuming, but is too often mixed with other fall-blooming crops such as goldenrod, some asters, and boneset.

TEASEL 📷 —*Diapsacus sylvestirs:* (shown at left) In North America from Canada to North Carolina and west to at least Missouri, this blue-flowered perennial grows abundantly along roadsides and in abandoned fields. Popular with bees, the light, delicate honey is too often mixed with other midsummer blossoms to be identified. The spikes on the flowers occasionally entrap visiting honey bees.

GOLDEN HONEY PLANT 📷 —*Actinomeris alternifolia:* (shown at right) Found along stream edges and other damp places, this fall-blooming plant is more important than typically given credit for. The light amber to amber honey, when isolated, is pleasant and aromatic, somewhat like its crushed leaves. It can be counted on as a reliable fall producer because it grows only where soil water is abundant year to year. If you find this plant growing nearby, move colonies to take advantage of this nectar source.

GOLDENROD 📷 —*Solidago* spp.: (shown at right) There are many species of goldenrod in North America, Europe, and elsewhere. They begin blossoming in July and continue until frost. The honey varies from strong, bitter, and thick, to a smooth, silky, butterscotch-like finish that is highly prized.

ASTERS—*Aster* spp.: Asters are generally fall-blooming plants, during and after the time when goldenrod flowers. The honey is strong, dark, and most often left for the bees because harvest is over before asters bloom. However, this is becoming a debatable issue as mite treatments are interfering with many of the late summer and early fall honey flows, particularly late goldenrods and asters. Evaluate your pest control programs carefully, but at the same time take advantage of these specific crops.

Golden Honey Plant
Actinomeris alternifolia

Goldenrod
Solidago spp.

Redbud
Cercis canadensis

Cherry
Prunus **spp.**

Wild Shrubs

From a beekeeper's perspective, flowering shrubs offer a more permanent source of nectar and pollen for colonies than common wildflowers do. Shrubs tend toward second or third generation in the growth of a landscape, whereas wildflowers tend toward filling local and temporary niches. Fireweed, for instance, is a pioneer plant, colonizing open spaces created by fires or logging. When the next group of plants comes along, such as seeding trees and understory shrubs, they shade out the pioneers, which soon disappear. The shrubs then tend to last longer and produce more until they, too, succumb to the shade and competition of the taller and hardier trees that eventually fill the space.

WILLOWS—*Salix* spp.: There are dozens of willow species that are shrubs, vines, and small trees both in North America and throughout the world. Like their more robust cousins, there are both male and female varieties and early spring pollen is valuable for brood rearing. The honey from these, however, tends toward light and mild, but with a slightly acidic bite as an aftertaste, likely due to their high moisture environments.

REDBUD ◙—*Cercis canadensis:* (shown at left) One of the earliest shrubs or small trees to bloom, the pink-red blossoms show before the leaves. Especially important in the central to mid-south sections of the United States for both buildup and surplus crop. The mild, light-coloured honey is cloverlike in flavor.

PRIVET—*Ligustrum* spp.: Common privet is found in the northeastern parts of the United States, where it is often used as a hedge but has escaped and achieved shrub status. This species produces a strong, foul-tasting honey, and even a little in a super can ruin a whole tank full of otherwise good honey. Meanwhile, its cousin in the south, Chinese privet, produces a light, mild, and very tasty crop. Midspring blooms are usually mixed with other crops.

CHERRY ◙—*Prunus* spp.: (shown at left) Several species of this plant exist: black, choke, cultivated sweet, or sour. All are early bloomers, all but the sour need cross-pollination, and all produce a dark, sometimes reddish honey with a bitter flavour.

GALLBERRY —*Ilex glabra:* (shown at right) Common in the southeastern United States along the Gulf of Mexico, gallberry produces a light amber honey that is particularly fragrant. A staple honey crop in that region, it is quite easy to harvest nearly pure crops. Granulation is especially slow.

MESQUITE —*Prosopis* spp.: (shown at right) Common in the North American desert southwest, but moving north and east, this mid- to late spring bloom produces an exquisitely mild, light honey that is easily isolated and harvested nearly pure.

SAW PALMETTO —*Serenoa repens:* (shown at right) Found in the United States from North Carolina to Florida and west to Texas along the Gulf Coast, this common, semitropical summer-blooming plant produces a good to generous nectar crop even in dry years, as it is accustomed to sandy soil and dry spells. The rich, yellow honey is mild, pleasant, and fragrant.

SUMAC —*Rhus* spp.: (shown at right) A multitude of varieties of this plant populate much of the eastern United States and produce much of the honey crop there. They mostly bloom in mid- to late spring. The typically dark, strong, and regionally favoured honey borders on unpleasant for many, but it is popular where it grows. The dried seedpods from the female plants serve as good smoker fuel when harvested in the fall and hung to dry.

AMERICAN HOLLY—*Ilex opaca:* This traditional Christmas plant produces a fair crop of honey in May and June in the east central United States. The honey is medium-flavoured and amber to dark amber and locally favoured, but not common out of the region.

HAWTHORN—*Crataegus* spp.: There are several species of this thorny shrub that bees visit, primarily for nectar. Found mostly in abandoned and old fields and pastures, it was once used for fencing. The resulting berries make good wildlife food, and the dense foliage makes good cover.

Gallberry
Ilex glabra

Mesquite
Prosopis spp.

Saw Palmetto
Serenoa repens

Sumac
Rhus spp.

Honeysuckle
Lonicera **spp.**

Snowberry
Symphoricarpos **spp.**

HONEYSUCKLE 📷 —*Lonicera* spp.: (shown at left) This introduced, escaped shrub is strikingly overlooked as a major honey plant. An incredible nectar producer when it blooms in early summer, this plant produces honey that is light, mild, and flavourful, with an aftertaste similar to the fragrant aroma of a blooming flower. The delicate blooms can be subject to drying winds or heavy rainfall and their bloom dates are shortened considerably. An escape from its imported ornamental beginnings, this plant is responsible for more "wildflower" honey than almost any beekeeper gives it credit for.

MANZANITA—*Arctostaphylos* spp.: Primarily a U.S. West Coast honey plant, it is found in the Rockies and sometimes farther east. Blooms early; some species begin in January, through midsummer for later bloomers. The honey is white to light amber, but some say the flavour is a touch bitter.

SERVICEBERRY 📷 —*Amelanchier* spp.: (shown at right) Used extensively as an ornamental, this small tree has escaped and become part of the landscape. If the weather is warm enough, the flowers can be very fragrant, but it usually blooms too early in the spring. Honey is light amber, mild, and just a little fruity.

SNOWBERRY 📷 —*Symphoricarpos* spp.: (shown at left) Another escape, this low-growing shrub is found mostly in the eastern United States and blooms summer to fall. The berries are good food for wildlife, and the honey is delightfully fragrant and mild.

SWEET PEPPERBUSH—*Clethra ainfolia* and spp.: Found mostly in the eastern United States. Blooms from mid- to late summer with long clusters of white, very fragrant flowers. Light to very light honey is fragrant, profuse, and perfume sweet.

SPRING TITI—*Cliftonia monophylla:* An important honey plant in the southeastern United States, it blooms early and is a primary spring buildup plant, but surplus is easy to get if the bees are ready. Honey is strong and medium dark.

SUMMER TITI—*Cyrilla racemiflora:* Another plant that grows in the southeastern and Gulf Coast areas of the United States, summer titi blooms later in the season and favours moist and swampy areas but can be found on higher ground. Honey is strong, dark, and regionally favoured.

CATCLAW—*Acacia* spp.: In the United States from Texas to California, this desert plant blooms in early spring and then again in the summer. The honey resembles mesquite in that it is water-white, mild, and slightly cloverlike in flavour.

RUSSIAN OLIVE 📷—*Elaeagnus augustifolia:* (shown at right) A spring-blooming ornamental that has escaped and become established nearly everywhere except desert areas. An import, it was used early on in the Midwest by conservation reserve programs as a windbreak in prairies and spread from there. An incredibly fragrant, prolific nectar and pollen producer, this shrub is overlooked in importance and contribution. The honey is mild, sweet, and extremely fragrant.

VITEX—*Vitex negundo:* A small to medium shrub, this ornamental has escaped and become established in the warmer areas of the United States. With adequate moisture, it blooms in summer to late summer. The honey is light and resembles sweet clover in colour.

SALT CEDAR—*Tamarix pentandra:* This arid-climate-loving but river-hugging invasive species has made a name for itself as a problem along waterways in the western and southwestern United States. The summer bloom, though, helps bees when nothing else is available. The honey is not table grade, but beekeepers like it because it makes excellent bee food when nectar stores are otherwise low.

Russian Olive
Elaeagnus augustifolia

Serviceberry
Amelanchier spp.

True Artisan Honey

More often than not, successfully harvesting a nearly pure varietal honey is impossible. There are too many variables involved in honey bee colony management: the weather, the dynamics of the plant community your bees are subject to, and just plain luck. However, if you are aware of which plants your bees have collected their crops from and when they collected them, and you have been able to isolate these crops to some degree, you will end up with a true artisan honey—it may be a blend of black locust and willow, or sunflower and soybean, or goldenrod and golden honey plant—but it will be a truly unique blend that can be marketed as such.

But negotiating the collection of these fine products is only part of the task. Once safely stored by the bees, it must be carefully removed, separated from the combs the bees have stored it in, strained, and stored—all without damaging the delicate aromas and flavours of the various and many ingredients contained therein. How to accomplish this follows.

Herbs and Mints

These few plants could easily be placed in other groups, but they stand out for the quality and quantity of the honey they produce. Because mature stands are often grown as harvestable crops, they tend to be quite populous and are able to produce nearly pure crops of varietal honey. If this is your goal, locations featuring these plants should be sought out.

THYME 📷 —*Thymus serpyllum:* (shown at right) Once common in pastures, it is far less abundant now because pastures are disappearing in favor of houses and parking lots. The honey is strong, minty, and medium in colour, and is easily identified by the fragrance. Blooms mid- to late summer.

LAVENDER 📷 —*Lavandula officinalis:* (shown at right) Common in home gardens, it is grown in some places as an oil crop for perfumes, commercial scents, soaps, and cosmetics. Midsummer bloom produces a distinct, though not overly strong, flavoured honey.

BORAGE—*Borago officinalis:* Common in small plantings in gardens, it is a commercial crop in parts of Canada, where it is grown for its seeds, which are crushed for oil. It blooms midsummer, and the honey is medium to dark but not as strong as one would suppose looking at the colour. A fair amount of secrecy surrounds this plant, as beekeepers are reluctant to share such a prize crop.

SAFFLOWER 📷 —*Carthamnus tinctorius:* (shown at right) Grown for oil primarily in the western United States, this midsummer bloomer produces an abundant crop of richly flavoured, darkish honey, occasionally with a greenish tint. Said to be a very reliable producer, no matter the weather or the rainfall.

Thyme
Thymus serpyllum

Lavender
Lavandula officinalis

Safflower
Carthamnus tinctorius

Chapter 3

The Honey Harvest

TO KNOW WHEN HONEY IS READY to be removed from a colony, you also have to know when it is not. That may sound simplistic, but too often beekeepers harvest their crop when it is most convenient for them rather than when the honey is in its peak condition.

WHEN IS IT READY?

To know when honey is ready to harvest (and when it is not), you need to understand the fundamentals of honey—how honey comes to be—starting with the mother of all honey, floral nectar. This product is manufactured by plants and distributed through their flowers, or perhaps the extra floral nectaries that some plants have in addition to or in exclusion of flowers. Plus, there's a curious product called honeydew, an excreted by-product of plant-juice-sucking insects that is collected by honey bees, then converted into a honeylike product. We'll deal with this unique and special product later.

The basic chemistry of floral nectar and honeydew is not quite as simple as you might suppose. Moreover, all nectars are not created equal, meaning the honeys made from them are not equal either.

Understanding the chemistry and the chemical transformations that occur after nectar or honeydew is collected by a honey bee is important so you can intelligently discuss this with your customers if you sell your honey. But more significantly, it is important for you to be aware of these principles so you understand the reasoning behind the techniques suggested for harvesting, extracting, straining, and storing your crop. These include the basics of sugar transformation that take place after nectar is collected and returned to the hive, the actions that occur after the nectar has been accepted and then stored in the hive, what can happen to not-yet honey once it is stored, and what can happen to the finished product in the hive after it has been capped and finished. This can be a critical time in the life of a product as delicate as this supersaturated sugar solution.

HONEYDEW

Honeydew begins as a phloem material from a plant that has been ingested and digested by sap-sucking insects, usually aphids or scale insects. It is excreted directly onto leaf surfaces and collected by honey bees and other insects. It is usually produced by oak, some pines, some sycamores, and basswood trees. It is produced more often during hot, dry weather, but that is not always the case. Honeydew contains enzymes, proteins, and other chemicals from both the plant and the insect, and a host of minor sugars. It is generally characterized as a poor-quality honey bee food, but it has a strong following by some and is often called tree honey or pine honey. The flavour can be strong and pungent, which is usually associated with pines as the source, or generally dark but not distinctive when coming from other sources. It usually has dust and sooty mold spores in it.

HIVE HINTS

Taste the Aroma

Leave your smoker on top of a nearby hive or a short distance downwind so it doesn't interfere. With a veil on and using the hive tool as gently as possible, remove the outer cover, then the inner cover (if there is one). A gentle touch avoids overly arousing the guard bees, though some will begin to pay attention. Allowing them to generate just a touch of alarm pheromone will invite you into the aromatic experience. Once the cover has been removed, slowly lower your face toward the exposed top bars of the frames. Get as close as possible without the veil pressing against your face. Draw a long, deep breath through your nose. Try to identify as many, if any, aromas as you can. Some are sharp and minty, others butterscotch-like smooth, and some will be maple-syrupy sweet, while still others will be fruity and fresh. The aroma of fresh-mown hay may be fresh pollen, and the lemony smell is likely Nasonov pheromone. It is a heady experience. Try this aroma test throughout the season and learn the different smells. With practice, you will know what the bees are foraging on simply by the collective aromas of the colony.

Honey comb cells partially filled with honey may or may not be ready to harvest. Uncapped cells usually indicate that the honey is unripe and not ready for harvest. Careful beekeepers avoid harvesting frames containing unripe honey.

Honey is a delicate blend of many different sugars, aromas (derived mainly from volatile ketones and aldehydes), enzymes, minerals, salts, acids, proteins, trace elements, amino acids, pigment and flavour compounds, vitamins, fats, lipids, and still more as-of-yet-unidentified substances. It is generally acknowledged that there are as many as 200 and probably more substances known to be present in most honeys. It is a complex, complicated, truly unique work of botany, biology, science, art, and possibly God (or gods). It should never be abused, it should never be intentionally mishandled, and it should never, ever be taken for granted.

Because of this complexity, the balance of its composite parts, and its fragile nature, extremes in processing techniques and storage conditions—especially extremes in heat—will cause changes in honey... and it's never for the better when honey is exposed to these environmental insults.

Factors other than heat that can and will change honey. Include:

- Providing high-fructose corn or regular sugar syrup in the hive when honey supers are on the hive (leading to unintentional honey adulteration)

- Having chemical agents used for pest and disease eradication in the hive when honey supers are on, in unprocessed full, but stored, supers, or in storage areas after extracting

- Allowing brood to be harvested and processed with the honey, lending additional and unwanted biological substances to the finished product

- Not keeping the storage area free from pests that can, and in all likelihood will, damage unprocessed honey, such as small hive beetles, wax moths, or rodents

- Blending two or more honeys together to mask the disagreeable taste of one with the more agreeable taste of the other

- Blending a light-colored honey with a dark honey so as to lighten the dark variety

- Blending a too-high moisture honey with one that's below the fermenting limit to reduce the likelihood of fermentation

Any of these activities will only harm the final product and are unacceptable beekeeping practices when producing high-quality varietal honey. And frankly, allowing any of these transgressions to occur borders on felonious culinary assault.

So let's begin at the beginning.

Nectar Conversion

Nectar is a sugary solution plants produce and make available to insects and other animals. Nectar is generally produced near the nectaries in flowers or in other locations on the plant. Simply put, nectar is a bribe to get these visitors' attention. Most flowers are constructed such that when insects, birds, bats, lizards, ants, or other visitors dip into the nectar for a meal, they inadvertently contact the pollen-producing parts of the flower and in the process collect small amounts of pollen on their bodies. Pollen, as you know, is the male partner in this sexual dance of seed production. When visiting other flowers of the same species, these accidental passengers—pollen grains—are transferred to the receptive female parts of the flower, and pollination and seed set happily occur. Flowers are generally a plant's attention-getting device that lets pollinators know where the nectar and pollen are, and the pollen and nectar that are found in flowers are the rewards pollinators get for that attention.

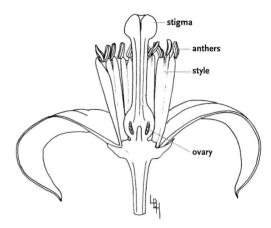

A longitudinal section through a Washington Navel Orange blossom, showing the important floral structures.

A partially dissected citrus blossom. At the base of the style is the ovule, where a fruit will develop if the flower is fertilized. Below the ovule is the nectary tissue, with obvious nectar present. When an insect such as the honey bee pushes into the flower to reach the nectar reward, she will brush against the anthers, picking up pollen grains in the process. When she next visits a flower, some of these pollen grains will be brushed off from our forager onto a receptive stigma and fertilization will occur ... and eventually orange juice happens.

Nectar is the raw product of honey. It contains the aromas, flavours, pigments, and many sugars ultimately found in honey. The predominant sugar found in nectar is sucrose, essentially common table sugar, dissolved in the watery plant sap. Sucrose is a fairly complex twelve carbon chain sugar that cannot be digested by honey bees, or humans for that matter. So to convert it to a digestible form, honey bees must transform it in a process called ripening.

Nectar collected from flowers (or extra floral nectaries, or honeydew from plant surfaces) by a honey bee forager is sucked up through her proboscis (the three-part, strawlike tongue). It enters the esophagus, passes through, and is then held in the far end of this tube in a chamber known as the *ventriculus*, but most commonly called the crop or honey stomach (an enlarged section of the esophagus designed to hold surplus nectar). Think of the crop much like the storage area a pelican has in its beak, where it can hold captured food until needed to feed its young.

At this point in the process, nectar is on average about 80 percent water and 20 percent sugars (mostly sucrose, but other sugars are present in varying amounts). The ratio of water to sugar may be as much as 50:50 to as little as 95:5. Obviously, honey bees prefer to gather nectar with as much sugar dissolved in the plant sap as possible; the richer the nectar, the more efficient the trip the forager makes. These ratios vary by plant species, sometimes considerably, but that 80:20 ratio is a good benchmark. When collecting the nectar, the forager adds a bit of saliva to dilute it slightly so it easily passes to the crop.

Though the proboscis needs to be airtight so it can suck liquid nectar, there is an area near the top where the three individual parts can be separated and the proboscis opened. After our forager returns to the hive, she opens that small slit and releases some of the nectar so house bees, scouts, and other foragers can taste the bounty. Eventually all of it is released to house bees, and even passed from bee to bee in the colony before the active ripening process—turning nectar into honey—begins, and our forager can return to the field.

A nectar-laden forager offers some of her bounty to a house bee to taste and analyze.

If the house bee takes the nectar, she will find a quiet place and begin the evaporation process. She does this by first adding enzymes—glucose oxidase and invertase—in a little bit of saliva she adds to the nectar. Then she extends a bubble of the nectar, exposing it to the warm, dry hive air. As the water evaporates, she adds more nectar and continues until she is needed elsewhere.

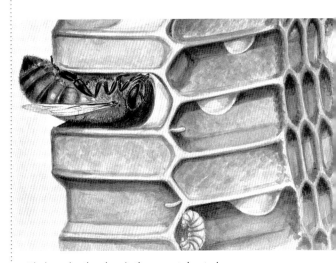

The house bee then deposits the evaporated nectar in an empty cell or one containing an egg or young larva to finish the dehydration and transformation process. When complete, or nearly so, the drop will be moved to a cell containing just honey.

CRYSTALLIZATION

....

The ratio of the glucose and fructose in nectar varies by the source of the nectar and can range from a low of about 1:3 to around 1:1. These two sugars amount to 60 to 90 percent of the sugars of honey's total carbohydrates. The remaining sugars are created when the enzymes are added, or were in the original nectar and are complex mono- and disaccharides that add subtle flavours, aromas, and body to the finished honey.

This ratio is important in the crystallization process honey almost always goes through. Recall that honey is a supersaturated solution—that is, a larger amount of sugar can be dissolved in water when the water is at the very warm internal colony temperature of 95°F (35°C) or so than would be under conditions at temperatures lower than this internal temperature. At this temperature a homeostatic condition is reached where the sugar and water are comfortable with each other. When honey is moved from this warm colony environment to a cooler environment, say, room temperature, or about 72°F (22°C), a condition will again be reached where the glucose and water are comfortable, but different, and some of the glucose will come out of solution and form crystals. Glucose is the least stable of the two sugars and comes out of solution the fastest when the temperature changes or the amount of water in the solution changes, or some other event triggers crystal formation.

Honey that has more glucose than fructose will tend to crystallize faster than honeys with more fructose than glucose. Moreover, when glucose does crystallize, it frees up the water that was holding it, which raises the moisture content of the remaining solution. Increased moisture can allow the remaining honey to ferment because the yeasts that are endemic in all honey now have an environment that allows growth.

Some plants normally produce nectars rich in glucose and are known for their propensity to crystallize rapidly. Almost all of the nectars originating from the Brassica family of plants—the mustards, canola, rape—are the champions at crystallization speed. Beekeepers who move bees to canola fields for these abundant crops know this, or learn it quickly and transport and extract those honeys as soon as possible—within a day or two at the most. Crystallized honey in the comb is a disaster because it is all but impossible to extract without damaging the integrity of the original honey and its comb.

Meanwhile, other plants produce nectars rich in fructose, and these are reputed to never crystallize. Chief among these is nectar from the tupelos in the southeastern part of the United States. (See chapter 2 for more information on tupelos and Brassicas).

EVAPORATING NECTAR

1. The mouthparts of our honey bee at rest with her proboscis folded down.

2. When she begins to make her nectar bubble, she extends her proboscis and ...

3. ... expels a small drop of nectar.

4. To increase the rate of evaporation, she rolls the drop and increases the amount of nectar until ...

5. ... the drop becomes quite large. A larger surface area enhances evaporation even more.

6. When the nectar has evaporated to about half or so, our bee adds more nectar and continues the process. She will continue until she has reduced most of the nectar in her crop, or she is summoned to unload foragers returning with more nectar.

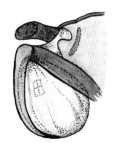

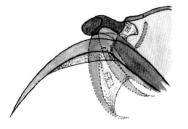

Enzyme Addition

The first step in the ripening process is when a house bee that ends up with this booty immediately adds the enzyme invertase to the now somewhat-diluted nectar. The addition of this enzyme begins the chemical conversion of the twelve-carbon sucrose sugar molecules into mostly two, six-carbon sugar molecules—glucose and fructose (see Crystallization on page 56). This enzyme is produced in the worker's hypopharyngeal glands and is added to the saliva she adds to the nectar. Other sugars are created in this process, but glucose and fructose are the most abundant. Glucose oxidase, another enzyme, is added to the nectar at this time also. This important compound reacts with some of the glucose sugar that's now in the mix, which allows water from additional saliva to be added, and gluconic acid and hydrogen peroxide are formed. Gluconic acid adds to the acidic property of honey, which aids in its chemical preservation, while the hydrogen peroxide has antibacterial properties that keep honey safe from biological harm while in storage.

Nectar Transfer

The speed with which the nectar is transferred from a returning forager to a house bee depends on the need for nectar in the colony, the quality of the nectar, and the number of available house bees in the colony at the time. If there's a strong need for nectar, the quality is good, and there are ample house bees, the returning forager will be unloaded quickly and be off looking for more. The house bee then adds a tiny bit of saliva and her enzymes to the nectar, and, if time permits, immediately begins an active dehydration process.

To do this, she allows a drop of this now-changed nectar to form between her wide-spread mandibles and her proboscis. This exposes the drop to the warm, dry air in the colony, and some of the moisture dehydrates. After holding this drop for a few minutes, she will add more nectar from her crop and continue the process.

Eventually, she'll place this drop on the side of a cell near the brood nest, or as far away as a honey super; the cell could be empty, have an egg, or even have a small larva in it. There the drop will stay for several hours, exposed still to the dehydrating atmosphere of the colony. As the drop loses even more moisture, it will be moved and added to other partially dehydrated drops in cells half to three-quarters full. There they stay for a day to several days, until the moisture content drops to below that magic 18 percent mark. When a cell is partially filled, say only a quarter to half full, the nectar will dehydrate faster than nectar in cells that are fuller. Twice as fast, in fact, from cells that are a quarter full as compared to cells that are completely full. This alone should be enough incentive to beekeepers to keep more than enough room available in colonies during a strong honey flow. Not only for the honey to come, but also for the nectar that it comes from!

THE HONEY FLOW IS ON

....

The bees are great at telling you there is a honey flow on if you've managed to miss all the other signs given off by the plants, the weather, and growing degree days. One of the very first things they do when they have incoming carbohydrate is begin making wax. If there's undrawn foundation, they'll begin to draw it out so they have places to store nectar and honey. If the foundation is already drawn but has been previously uncapped and the edges are uneven and rough, they will smooth them out and even them up. If even this has been done, those wax-producing house bees will place a thin film of brand-new beeswax on the top and sides of top bars. Anywhere they can, bees will begin to put new wax.

This new wax is beautifully white and will contrast nicely with older comb or with propolis-covered top bars, and it will certainly be noticeable as it is added to foundation to make new cells. Look for the obvious sign of a brand-new honey flow: brand-new beeswax on almost anything and everything inside.

House bees continue to test these drops to see how much moisture remains, and when ready may move the now-honey liquid to a cell that will eventually be full of honey. These cells may be near the brood nest in the frame corners outside the pollen ring that surrounds the brood nest, or if these are full, they'll be moved to the surplus honey storage area, generally in the honey supers above the brood nest. Some of this honey goes to cover cells already partially filled with pollen. This covering prolongs the active life and quality of the stored pollen. It may be capped, but sometimes it remains uncapped for the duration of storage.

When honey cells are full and the moisture content of the honey in the cell has fallen below 18 percent, the cell will be sealed (or capped) with a wax covering by wax-producing house bees. This cap is somewhat porous, however, allowing some air and water vapor exchange to occur as they change within the colony.

Cap Placement

The genetics of the colony's bees dictate how that wax cap is placed on that cell of honey. Some will lay that wax cap directly on the surface of the honey. This causes the wax to look as though it is wet (these are called wet cappings)—imagine laying a paper towel on a puddle of water on a tabletop for the same effect. Others place that wax cap over the cell, leaving a slight air space between the surface of the honey and the underside of the wax cap. When this happens, the wax remains white(ish). These are called dry cappings. You can see that the only difference is the location of the wax cap over the honey, but there is no difference in the finished product—the honey—only in the appearance of the comb's surface before it is processed further. Once honey cells have been capped, the honey is ready to be removed from the hive, and this is the next step in harvesting varietal and artisan honey.

Environmental Factors

But wait. This entire process is affected by the external environment the colony is in. For instance, colonies that are in more shade than sun may take longer to dehydrate nectar because of the potential temperature and humidity differences that exist. Cooler air entering a colony certainly holds less moisture than the warm, dry air, and the warmer the air, the faster the honey inside will dehydrate. Colonies that have ventilation problems—small openings, or openings at only the bottom or the top—will have slower air exchange than colonies with average ventilation.

The frame on the top has honey with dry cappings. The wax covering is placed by the bees such that there is an air space between the wax and the honey. The frame on the bottom has honey with wet cappings. The wax covering is placed by the bees such that it rests directly on the honey below.

☞

Carniolians (*Apis carnica*) are known for producing beautiful, snow white cappings. If you are interested in comb honey, these bees are for you.

☞

Italian honey bees (*Apis ligustica*) are also known for producing dry, white cappings. The only race that consistently does not produce white capping are Caucasians (*Apis caucasica*).

☞

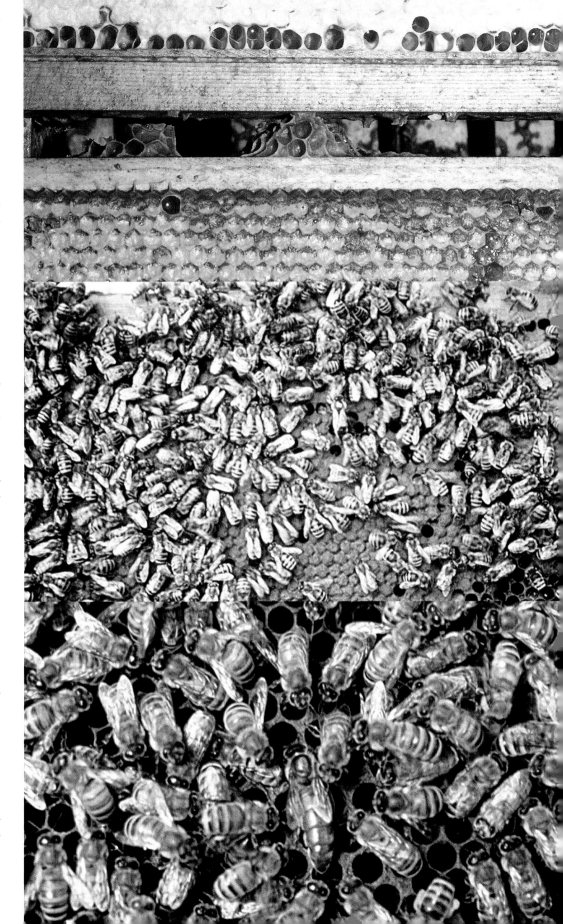

OLD COMB

·····································

It has been shown many times that when a light honey is stored in comb that is dark from old age—brood inhabitation or for whatever reason—the honey will be colour enhanced, that means darkened, by the materials in that comb that made it dark in the first place. These materials might include propolis, old cocoons, stuck pupal frass, or even the remainders of darker honey that was stored there previously.

As a result, the honey will be darker when harvested than surrounding honey from the same source that is stored in newer, less-stained comb. There is no way to lighten this honey, so the only way to avoid this is to continually recycle combs with brood cell history out of honey production and into brood production and eventually into the wax melter to be recycled. If you use mediums or shallows for honey production and deeps for brood production, you don't have that option, so they should go to the melter right away. It's still clean wax and it will make good candles. Though reusing old comb does enable you to get more mileage from those frames, keeping old comb out of the hive is always a good idea. And darkening honey is a crime.

This situation points to one of the benefits of having screened bottom boards on your colonies: increased ventilation and quicker drying time for the nectar your bees are bringing in. Moreover, this increased passive ventilation—that is, air movement not instigated by the bees themselves—allows more bees to be doing other tasks in the hive. Few, if any, bees will be fanning air to move it through the colony, enabling them to be attending to other tasks, such as nectar collection.

Eventually, the available space in a colony will be filled if there is adequate nectar in the environment, enough bees to collect it, and your colony continues to operate within a normal season.

Traditionally, the procedure has been not to remove frames of honey to extract that are not capped. This is a good rule of thumb. The honey in cells not yet capped is almost always at a moisture content above that magic 18 percent. Recall that honey is a supersaturated sugar solution, and when the moisture content is below 18 percent, there isn't enough water present to allow the yeasts that are endemic in the colony to grow, causing the nectar solution to ferment.

Rarely nectar, or even partially dehydrated nectar, will ferment in the cells before it is completely reduced. Once nectar or honey ferments in the comb, it appears as bubbly honey in uncapped cells. This isn't common but it does happen, some seasons more than others and in some places more than others. If honey does ferment, it is not fit for human consumption (if you suspect you have some, go ahead and taste it—you will agree), and generally even the bees don't like it.

Review

So far we've covered several areas that can be seen as roadblocks to producing as much honey as possible, or as much *good* honey as possible. Of course, there are always natural limits to how much honey your bees will produce; they could encounter inclement weather, pests or diseases that reduce the efficiency of your foragers, an agricultural pesticide poisoning, or the sudden and unexpected removal of the crops your bees were harvesting.

But many obstacles can be avoided, and too often beekeepers are their own worst enemies when it comes to what they could have done better for their bees and their crop. Because these principles are so fundamental and so critical to producing the most and the best honey possible in a given situation and location, they should be reviewed, again, to make sure they receive the amount of attention they deserve.

Honey Don'ts ...

DON'T contaminate your crop with honey bee food, such as sugar syrup, or pest control chemicals.

DON'T harvest brood.

DON'T allow pests in the honey supers after the honey is harvested.

DON'T mix light and dark honeys when extracting (we'll look at this in detail later).

DON'T blend honeys after extracting to hide colour, moisture, or taste.

DON'T harvest uncapped honey.

And Honey Dos ...

DO have enough healthy foragers *before* the honey flow to harvest a major honey flow.

DO have enough healthy house bees *before* a honey flow to process a major honey flow.

DO have more than enough temporary storage space in a colony *before* a major honey flow begins.

DO have more than enough permanent storage space in a colony *before* a major honey flow begins.

DO have excellent colony ventilation.

DO provide good colony locations for internal honey reduction.

DO make sure the bees in the colony remain healthy, avoiding diseases, pests, and, most important, mites.

HONEY REMOVAL HISTORY

Very little has changed in the world of separating bees from their honey in the last one hundred years or so. Way back, feral bees were driven out of their nests and off combs (this is before hives were used) by using excessive amounts of smoke. Though effective, the technique lacked organisation, and when skeps came into being, killing the bees so the honey could be removed seemed a better way to go. Bee brushes go back decades and served as the mainstay for quite some time after movable frame hives came to be, because killing bees was no longer necessary (the bees were pleased with that innovation). Plus, when the bees were brushed away from the frame, they could then rain back down on the top of the colony or at the front door. Bee escapes have been around nearly as long as movable frame hives that used supers, and though the early metal devices have given way to plastic, other designs have become popular (even so, the oval hole in the inner cover exists solely for that reason).

Fume boards have been used for decades, though the repellents used on them have evolved from dangerous acids (they used to be called acid boards because carbolic acid was the first "fume" to be used) to the trendy herbal oils of today. Blowers, too, have been used, though not for quite as long. They have evolved from the large, heavy, cumbersome, noisy, not-too-reliable early models to the (nearly) all plastic, lightweight, dependable leaf blowers commonly used now. Though the techniques haven't changed much, the tools we use have made the task safer, easier, and faster for us, and certainly easier on our bees.

MARKING YOUR FIELD BEES

You may think this is a complete waste of time, but before you dismiss it completely, I challenge you to try chasing your bees from your hive to the crop they are foraging on, and then from the crop they are foraging on back to your hive. It's not as difficult as you may think.

First, let's find out where your bees are going. To begin, refer back to your map of what's growing where you developed previously (or are going to develop now). Use Google Earth or a similar device to locate your apiary, and then outline the surrounding area for at least 2 miles (3.2 km) away. Driving and walking around will help, but an aerial view is indispensable, if at all possible.

Then, using the map, watch your bees when they leave the apiary to begin foraging in the morning. Follow them as far as you can, map in hand, if you can to where they go, but it is unlikely you will be able to keep them in sight. See where the map tells you they should be going; your map will tell you what's growing where, and you'll know, now, what should be blooming where the bees are.

Try this on a sunny day: While watching the hive entrance, see the bees rise and leave the front of the colony. Stand so the sun is behind you; you will see sunlight reflecting off their wings. They will turn quickly, bank, and circle to get their bearings before flying in a certain direction, or they may just fly away. Shift positions as needed so you can see where they go after turning, keeping the sun to your back. Be patient. You will soon be able to spot a flash of wing reflection for the nanosecond it exists, though, and be able to see in which direction they eventually head.

Check Your Map

Check your map and take a guess where they may be heading, knowing what's blooming in that particular direction. It may be a field of mustards, a stand of locust, or the city park. Venture out to that spot to see whether, indeed, your bees are there. Your bees? How do you know they are your bees? Why, you marked them. As foragers are on the landing board, sprinkle powdered sugar on them. This certainly makes it easier to see them when they leave, and easy to identify them when in the field.

Once you arrive at the location you think your bees should be visiting, sit down and wait. Watch the flowers that are blooming ... there are flowers blooming, right? If not, check your map again and seek out the alternative locations you marked. If there are bees visiting, first do this. Get the sun to your back again, so you are between it and the plants that are blooming, and do the sun thing again. Watch for the wing reflections bouncing off bees heading in your direction. Of course, if you're lucky, one of them will have powdered sugar on it. They should be coming in from the general direction of your hives. This works even if they are visiting blossoms up in trees. Watch as they arrive, and as they leave. This goes a long way in deciding whether they are from your colonies or not, but you can go one step further.

Sugar Shake

When you leave for your forage area, take a shaker full of powdered sugar with you. When you get there and see a bee on a blossom, cover her in sugar (if not already covered, which makes this a lot simpler). She will remove much of it before she flies home, but some will remain and be easily visible when she returns. Your job is to be there when she arrives. Of course, you have to powder a lot of bees and beat them home, but as the day wears on, if these are your bees, you will see them at the door, inside, back again in the flower patch, and sometimes just preening on the front porch, cleaning off this tasty marker you have made.

It will take many tries, and perhaps several different crops before you master this. It's never easy, even for the experts. Practice is fun, educational, and good exercise. It is surprisingly easy to see a powdered bee after some practice, even when they are sky-high. If they are visiting another yard's dandelions, practice will be curbed. Even so, getting close to where your bees are foraging is exciting, and using your map, your notes, and your plant knowledge in knowing what's blooming where is a skill few beekeepers ever master. Be proud.

Using a small mesh sifter or shaker, or another device, sprinkle powdered sugar on your bees as they are leaving the hive. This makes it easier to see them as they fly away, and identifies them when you see them in the field.

☞

Where the bees are ... may be a neighbor's flower garden ...

☞

... or a field of goldenrod alongside a railroad track.

☞

COAXING THE BEES

There will be a time when your varietal crop has been harvested by the bees, dehydrated, and capped and is ready to harvest. You need to act fast to ensure that what you are harvesting is what you want and that you aren't mixing crops, if that's possible. So at some point you will have completed the following tasks if you have been ahead of the curve and paying attention to all the details:

YOU'VE IDENTIFIED THE HONEY PLANTS growing in your forage area and the sequence of bloom dates. You have also calculated from previous data or located an existing source of growing degree day information so you can anticipate their bloom date exactly.

YOU HAVE TENDED TO and prepared your colonies so that you have a large population of healthy house bees and foragers well ahead of the anticipated bloom dates.

YOU HAVE PREPARED HONEY SUPERS well in advance of the anticipated honey flow, so you will have plenty of both temporary and permanent storage space for incoming nectar and honey storage.

YOU HAVE OBSERVED YOUR BEES visiting the crop you intended them to be visiting, when they should be visiting that crop (see Marking Your Field Bees on page 62).

YOU'VE OBSERVED YOUR MARKED BEES returning to your hives, so you know the nectar source they are visiting, that they are collecting that nectar source, and that they are returning to your hives with that nectar.

YOU'VE SEEN NECTAR BEING STORED, and you've seen honey being stored, and you've tasted the honey to confirm its origin (or you've learned what this honey tastes like, finally).

YOU'VE MONITORED THE GROWING DEGREE DAYS during bloom so you can anticipate the end of the bloom cycle for a particular crop or group of crops.

YOU'VE EXAMINED HONEY STORAGE AREAS and even the brood nest area for early signs of a honey flow (see The Honey Flow Is On on page 58).

YOU'VE EXAMINED THE HONEY STORAGE SUPERS for the ripening process and eventually the capping process.

YOU'VE ISOLATED THOSE SUPERS OF CAPPED HONEY and organised individual frames within the supers so each super is as full as possible of capped frames of the honey you have been monitoring.

Now it is time to remove the honey from the colony and get it extracted as soon as possible. Let's do that.

There are as many ways to separate honey bees from their honey as there are beekeepers, and all beekeepers have likely refined the way that works best for:

- The equipment they have

- The time they have

- The method of transportation from the apiary to the honey house that they have

- The distance they have to haul the honey and the place they will ultimately store and extract the honey

Each of these variables can present an obstacle to efficient honey removal and should be examined in some detail, both to look at the problems each situation presents and to find the solutions that are available, and perhaps even better options that exist. Far too many beekeepers spend 90 percent of their time preparing to make a honey crop, and only 10 percent of their time preparing to safely, hygienically, efficiently, and easily harvest, transport, store, extract, and finish their crop. This is unfortunate because too often what results is a poorly designed, inefficient, unsafe, and not-as-clean-as-it-could-be extraction facility.

Removing Honey Supers from Your Colonies

Once you have assessed your honey crop and decided that there is at least one and perhaps more supers that contain the honey variety you are interested in, that the honey has been ripened, and that each frame within the super is full, the super is ready to be removed from the colony and extracted.

Before you do any harvesting, a thorough examination of the colonies to be harvested should be undertaken. Frames with capped honey are easy to identify and are ready to remove and process when all of the cells are capped. If there are frames with uncapped honey or frames with brood in some of the cells, they should be exchanged for capped honey frames from other supers, if there are some and the honey is the same variety, or at least put back in the colony to finish maturing (brood) or to be capped (honey). If you can, try to keep supers or frames of like honey together for processing later—even if there are fewer than ten frames in the super when

Avoid the drippy frame by visiting your beeyard the day before harvest and separating supers. The bees will clean up any drips and seal over broken comb during the night, and you'll harvest clean, drip-free supers.

Separating honey supers the day before harvesting is always a good idea, if time and energy permit. Lifting and moving supers breaks propolis and burr comb seals and lets you know where bee space and factitious frames go astray. Breaking them apart a day before harvest is time-consuming and expensive, but it makes harvesting so much easier, cleaner, and less frustrating.

CHECKING THE MOISTURE

....

Checking the moisture content of your honey before you remove it from the colony is always a good idea but a task too seldom accomplished by beekeepers.

The common rule of thumb is that the bees know best when the honey has, in their judgment, reached that magic moisture content of just less than 18 percent and the cell is completely full of the now-honey solution; when the bees consider it finished, they cover it with newly made beeswax. This is the signal to beekeepers everywhere that the honey is "ripe" and can be safely harvested with no fear of fermentation later. This is generally a correct assumption, and if you never check further, you would hardly ever have a problem.

However, though not common, sometimes uncapped honey is at or below that magic mark and can be safely harvested, and sometimes capped honey is above that magic mark. Without checking first, you will be unpleasantly surprised to find fermented honey, or be kicking yourself for not harvesting perfectly safe honey. And you can't tell just by looking.

Using a Refractometer

If you don't already own a refractometer, get one. There are many models on the market; one common style is the traditional solids meter, which measures the solids in a liquid (and thus shows the percentage of the solution that's not solid). Not difficult to use, some models require calibration while others self-calibrate to accommodate temperature differences. These come in a range of prices, from fairly inexpensive imported plastic models to very expensive metal and glass models. Digital models are also available. These are less affected by temperature changes but do need some calibration adjustments. They are quite reliable, accurate, and durable, only requiring a battery to work fine in the field or honey house.

Checking the honey in the supers to be extracted the next day is insurance that what you will be doing won't cause problems down the road and will minimize surprises. When monitoring moisture content, careful beekeepers check a blended drop taken from inside and outside frames from every super when there is a history of damp honey, high humidity conditions, or recent rainy weather.

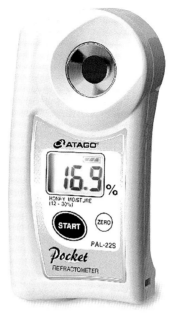

Digital refractometers are ideal for use in both the beeyard and the honey house. They require no calibration when the temperature changes; they are quick and easy to use, requiring only a drop of honey and a damp rag to remove the previous honey sample. They are very accurate, no matter the conditions they are used in. If there's a downside, it's the cost for this convenience; small operations may have difficulty justifying the expense when compared to a (slower) inexpensive glass and plastic manual model.

Uncapped honey can be above, at, or below the optimum moisture content, and you'll never know without testing with a refractometer. Still, the rule of thumb says not to harvest uncapped honey.

you are done exchanging these frames. Of course, this is easier when you have only a few colonies rather than hundreds, but once you begin compromising the quality of your honey, where do you stop? Take the extra time and work and keep the varieties separate if at all possible while in the field. You can be more exact once back in the extraction facility, but the more you do now, the less you will have to do later.

Once organised, make plans for the actual removal. If you have done this for a few years, you probably have a routine for removing, transporting, and storing supers, then uncapping and extracting the frames, and finally straining and storing the honey you now have. Nevertheless, here are some suggestions that may streamline your process and make harvesting varietal honey easier. They are techniques other beekeepers have found really can help.

MAKE SURE THE SPACE YOU WILL BRING YOUR SUPERS TO IS CLEAN, WARM, AND READY. You will know about how many supers you will bring home—do you have room for that many? Use drip boards—either upturned covers or specially made boards that catch any drips from the supers to keep honey off the floor.

VISIT THE BEEYARD AND ORGANISE THE SUPERS YOU PLAN TO HARVEST. Regardless of the technique you will use to separate the bees from the frames and supers, do this the day before you are to remove the supers. Remove frames with brood and frames with uncapped honey, and gather the frames to be harvested into one or more supers, moving the others to supers that will stay behind.

SEPARATE THE SUPERS TO BE HARVESTED FROM THE REST OF THE HIVE AND EACH OTHER BY BREAKING THE PROPOLIS SEAL. Simply tilt them up a few degrees. This breaks any burr comb and propolis seals between frame tops and bottoms, or between frames and box sides if other manipulations haven't done this already. Overnight, the bees will clean up the dripping honey from the broken comb but will not have time to remake it. This makes it much easier to remove the honey supers the next day and keeps honey drips that would have been spread everywhere later to a bare minimum.

If you do brush bees off frames for harvesting, use a soft-bristled bee brush and gently drop the bees at the front door so they don't get lost and will have somewhere to go.

If you use a Porter Bee Escape model for removing bees, make sure that all other entrances are sealed (masking tape works well for an overnight sealant), that there's no brood in the super, and that you have broken all the burr comb connections the day before. And, make sure the escapes are clear, no matter which escape device you use.

On the day of harvest (or better, the day before), plan ahead. Know where and how the honey supers will go once removed from the colony—in the back of your truck, in the trunk of your car, in a wagon, or simply carried to your house or garage. Whatever the means, make certain you have something to cover both the top and the bottom, so dust, dirt, bees, and debris don't get in. Remember: this is food; keep it clean. Drip boards in your truck, trunk, or wagon work well; keep drips to a minimum; keep dust, dirt, and bees out; and reduce any chance that a robbing situation will begin.

The Right Equipment

As an experienced beekeeper, you most likely have a technique you are comfortable with, but I offer some thoughts on the process that will, perhaps, make the job easier for you. As with everything else in beekeeping, your particular setup may be something that works just exactly right for you, but there may be something here that helps. Give bees a chance.

Brushes

Don't use a brush to remove bees from one frame at a time, even if it's the best way you know how. It infuriates the bees, damages the cappings, upsets the entire apiary, and takes forever. Well, not quite, but it is the most difficult way there is to do this and creates all manner of problems.

Bee Escapes

Bee escapes have a long history in bee removal. These are simply one-way doors—bees leave a honey super and can't return, or at least they can't return the same way they left. There are various escape devices, ranging from the traditional escape that fits in the inner cover hole to the multiple cone board to the triangle Canadian board. All work the same way and all have the same flaws no matter the number of escapes, the size of the corridor the bees must travel, or the cost of the device.

The theory, of course, is that an escape device is placed between the honey super you wish to remove and the rest of the colony below. It's put there toward evening, when the days are beginning to cool anyway, so that in the evening the bees will leave the honey super above the escape device and return to the brood nest below to form a loose cluster to keep each other and the brood warm.

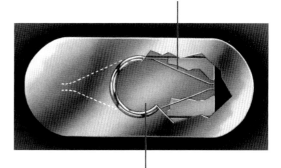

Bees exit through springs into the super below, and cannot return.

Bees enter here from the honey super you want to remove.

This shows how the Porter Bee Escape actually works. The two metal springs inside can be pushed apart by workers, and even drones if set correctly. The open space between the spring ends should be exactly $3/8$" (9.4 mm). If larger, workers can return. If smaller, drones will get trapped and block the escape from further use. To calibrate, slide these escapes apart and lay a pencil between the spring. A pencil is exactly $3/8$" (9.4 mm) in diameter, and you can pry apart or pinch together the ends to meet that exact width.

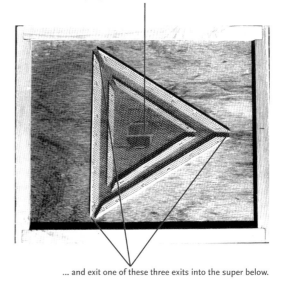

Bees exit the honey super above here ...

... and exit one of these three exits into the super below.

The bottom of the well-used triangle bee escape board is shown above. Bees leave the super above through the hole in the center and exit through one of the three points of the triangle. It is nearly impossible to find that entrance again and backtrack to the super above. The advantage of this is that the hole is large and there are three exit strategies available to choose from.

This works if you are harvesting later in the season when the weather is cooler. However, when the weather stays warm or you need to harvest before cooler weather arrives (almost a given when harvesting several times a season for special crops), getting bees to leave honey supers simply because you want them to can be a challenge. Moreover, unless you use a queen excluder, it is likely that there will be some brood in the honey super. That will keep bees from leaving altogether, so when you return the next morning, there will still be bees in the honey super.

Another consideration is that if you try this during the warmer part of the summer, the bees are not inclined to leave the honey supers because it's warm enough for them to stay where they are. Moreover, during the active brood-rearing season, again unless you use a queen excluder, there is likely to be some brood in that honey super. Same story ... the bees won't leave. This scenario applies to all harvests that occur during early and midseason. Escape boards can be difficult or impossible to use until later in the season when the weather turns cool. So what to use?

Using Fume Boards

Our first selection is to use a fume board to move bees out of the supers. There are several materials, or fumes, that you can use, and some are better than others. Most offensive is the Bee-Go product. It is efficient, fast, and so foul smelling that you will be sorry you ever encountered it. But it works. It works even on cool days with little sun. But it can leave an odour in the super and even in the honey. Stay away if at all possible.

A product closely related to Bee-Go is the same material with a cherry aroma added. It works equally well, but it also has a foul odour (though less so). It, too, works well in cool weather with little sun, but spill it on your boots or clothes, and you might as well burn them.

If you have hundreds of supers to move and not much time, both of these materials are effective and are the first choice of nearly all commercial beekeepers. But if you have a manageable number of supers to remove at any one harvest, there is a different product to use, called Bee-Quick. It's an all-natural, nontoxic blend of natural oils and herbal extracts (the exact ingredients are a secret) that to us humans smells just fine—like almonds, actually. But the bees don't like it as much as we do and will move away from it, but not quite

A typical fume board is essentially a box that is about 2" (5 cm) deep, with the external dimensions of a super. It's lined with an absorbent material such as cloth, felt, or cardboard, and has a dark top that absorbs the sun's warmth.

When using a fume board, place the appropriate amount of repellent on the underside absorbent pad, and then place the board on the colony slightly askew so the system is not overloaded. A smaller dose starts the bees moving, but not in panic mode, producing a slow, easy exit. Too much repellent will remove all the bees from the hive, not just the super. Leaving an open space will reduce the likelihood of this happening. After about 10 minutes or so, depending on the number of bees in the super, the temperature, the sunlight, the time of day, and the repellent you are using, you can move the board to completely cover the super. After another 5 to 10 minutes, the bees should be clear and you can remove the super.

as fast as they do from the other products. If the weather is warm and sunny, this is the stuff you want. It works in cool weather, too, though in cool and cloudy conditions it's slow and not quite as efficient. That's barely a downside. Spill it on your boots or beesuit, and it can be washed right out. The cost is reasonable but availability is sporadic.

Using Fume Boards

No matter the product of choice, the technique is simple and straightforward. Open the colony (don't smoke the front door first because you don't want bees moving up into the colony before you start) by removing the cover and inner cover, if you have one, using some smoke to begin moving the bees down into the super or brood box below the top honey super. Apply the correct amount of the repellent to the underside of the fume board (read the label), and set the fume board on the colony slightly crosswise, fume pad side down. Move on to the next colony and do the same. Then a third. By this time, most of the bees in the first colony will have moved down, but not all the way into the box below. Return to the first colony and turn the fume board so it completely covers the super. Move on to the second and third colonies, doing the same.

After you've covered the tops of three or four colonies, the bees in the first colony will most likely have moved completely out of the super, and it is ready to be removed. But be patient and move carefully while doing this. Maybe wait a few minutes between each colony for each maneuver. It used to be, when beekeepers were being taught this technique, they'd be told, "Be patient, have a smoke." That isn't politically correct anymore, but the timing still works. Ask yourself: do you want it right, or do you want it fast?

When you think it's ready, tip up the top super of the first colony, hinged on one edge so you can see the bottom of the frames of the super you have lifted. If there are bees still on the frames of that super, it's not ready to remove. Set it back down, replace the fume board, and wait until no bees remain in that between-super space. When they're gone, remove the super to the transport and cover it to keep bees and dust out. Use your drip tray if you have one, or an overturned cover to catch drips. This keeps any bees that are inside, inside, and it keeps other bees from robbing this now-unprotected honey. Begin a fourth colony with the equipment from the first colony, and move back to the second to get the super ... and so on until finished.

Using a Bee Blower

Next to fume boards, using a bee blower to separate bees from their honey is extremely popular. But, blowers are mechanical and can break down. Also, never use a blower on bees where they may be walked on, and always cover your supers when finished blowing.

There are several models of dedicated bee blowers on the market, and many beekeepers make their own.

Stage the supers such that when you blow bees out, they have a safe place to go. The back of the truck works well because you won't have to move the supers twice, but make sure you aren't walking on the bees when you carry additional supers to the truck.

Blow bees out of the super from the bottom first because the space between the top bars is usually wider and allows easier exit. Then turn the super around and blow from the top down so you get all the bees hiding inside.

Bees will accumulate on the truck bed surface, on the tops and sides of the supers. Keep them moving by blowing them off to the safe place below.

Some beekeepers will use a bee blower to remove any errant bees out of these supers before sealing them for the trip home. To do this, they stand the super on the narrow end on top of the colony they are clearing and simply blow any remaining bees out. Bees are blown away from the path back to the truck or car so they don't get walked on. Removing all of the bees from a super is a safety measure that is good business and keeps neighbor and family relations harmonious.

Bee Blowers

Bee blowers are used by many beekeepers to remove bees from supers. Blowers are efficient and fast, safe and thorough. You will bring home fewer bees when you use a blower than you will if you use any of the products just mentioned, no matter how thorough you are. However, you have to have a place in the beeyard to stage the super so the bees land safely and won't be walked on. That's first.

There will be lots of bees in the air when you use a blower, all over the place. This may not be a good choice for a beekeeper unless you have a large space to work in. Blowers are loud (even if you use a leaf blower with a modified nozzle to funnel the forced air between frames), and you will stir up the bees. If you are in an apiary in the country, they work wonderfully. But they have moving parts ... and moving parts have the bad habit of not moving when you are far from home, so the trip is wasted or made much longer for the repair time. How good are you at repairing small gasoline engines, in the field? And you have to have gasoline, or a gasoline/oil mix. Don't forget the fuel can.

To use a blower, find a way to situate the super so you can stand it on the narrow end and blow between the frames so the bees are expelled to a safe place. Place the super off the ground so that when done, you don't have to lift the super very far to the back of the truck or wherever it will eventually go for transport. Using one side of the back end of the truck certainly reduces multiple moves, but make sure you aren't walking on bees as you move supers from the hives to the truck. Bees blown out of a super are incredibly confused, and many will simply sit on the ground until they get their bearings and can fly again. They may stay there for an hour or more because they are that confused. Be careful not to crush any—never harm your bees.

Some bee supply companies manufacture and sell a collapsible device that, when unfolded, resembles a folding camp stool. The top of the "stool" is simply an open frame. A super is placed on the top of the stool and forced air is aimed down between the frames. The expelled bees exit the bottom of the super and are ricocheted off an angled canvas platform so they land in front of the stool on the ground (the beekeeper then is standing behind the stool). One flaw in this system is that you can't simply turn the super over to blow bees from the bottom up because the frames would fall out if you did so. Some beekeepers have made their

MODIFIED BEE BLOWERS

Dedicated bee blowers are generally small gasoline engines that propel fans, channeling air into a hose and through a narrow nozzle so air easily moves through the small space between frames. These engines are usually two- or four-cycle models and are generally reliable. Some beekeepers have modified home leaf blowers by purchasing or making a nozzle that does the same job. Sometimes this needs to be attached to the end of the leaf blower tube with duct tape or by other homemade means. Inventive beekeepers have constructed dozens of devices that accomplish blowing bees out of supers using vacuum cleaners, shop vacs, and even more curious inventions. The primary purpose is to blow air at a velocity fast enough to move bees out from between frames when they don't want to move; the blower must also be lightweight and easy to maneuver, and probably most important, it must be reliable when away from the home shop.

Super Sense

There is a safe, sane, and sanitary way to get your honey home. There should be a cover over the supers during transport, the supers should be belted together so they don't slide around and tip over during the drive. The drip board underneath should be built to accommodate a two-wheeled cart to unload supers down a large, well-supported plank incline.

Once you're home, unloading can take many forms. This beekeeper keeps covers and drip boards on the truck, and unloads supers onto a roller for easy transport right to his uncapping machine. There's a minimum of labor involved with this sort of setup.

own better-designed models using sturdier materials, and they work okay. As long as the bees go down and away from wherever people are walking, this is a sound practice. But it requires moving a super at least twice, increasing the workload in the field.

Removing Bees Safely

When working on a platform such as the back of a truck or the top of the colony next to the one being worked on, first blow between each frame from the underside of the super (the bottom bar side) to move bees out and away. The space between top bars is most often larger than the space between bottom bars, and bees are moved more easily from the bottom up. Once complete, spin the box 180 degrees and blow from the top bar side of the box out toward the bottom bar space. This dislodges bees stuck between the frames or on ridges or chunks of burr comb inside. Bees will congregate on the outside of the super and on the surface the super is resting on. Keep these bees moving by blowing them off onto the place where the other bees are. You may have to blow bees from both sides of the super a couple of times just to get them out, but by then most will be gone. Move the super to the transport area and cover it to keep bees and dust out.

No matter which technique you use—brush, fume board, or blower—try to remove honey between midmorning and midafternoon, which is when most of the foragers are away from home and you have the least problems with guards who are watching the front door for all those foraging foragers. This makes it easy on you and the residents of the colony.

And, no matter which technique you use, the rules for harvesting, transporting, and storing are the same. Honey is a food, and others will be eating this food. Treat it as such. Keep it covered during the harvest to keep robbing bees out (they will puncture cappings and the honey will drip out, making everything sticky and messy, right after you went to all the trouble to make sure that didn't happen), to keep dust and dirt out during your travels, and to make sure errant things don't fall into the super during and after transport. The key to being safe and clean is to keep the supers sealed while in the field with overturned covers, drip boards, or other types of protection. Covering keeps dust, dirt, and bees out. As beekeepers, we too often become cavalier about the cleanliness of our honey. Slow down, look at what you are doing, and do it better.

If you have been checking the moisture content during all of this, you know that the honey you are taking home is dry enough to extract without fear of later fermentation. So next comes the rest of the story: uncapping, extracting, straining, and storing.

Keep It Dry

A serious problem you will encounter during this adventure is finding supers with frames of uncapped or unripe honey. Checking the moisture content of the honey in these uncapped cells will tell you whether they are ready to harvest or not, but switching bad for good frames in the field is a time-consuming, difficult, and generally messy task. However, harvesting these not-ready-for-prime-time frames is simply asking for trouble later. High-moisture honey will ferment and destroy the quality of the crop you and your bees have worked hard to produce.

The same goes for finding frames that are partially filled with capped or uncapped brood. Bring these home and some of that brood will emerge, some will die, but worse, when it gets run through the uncapper and extractor, you end up with brood juice in your honey. Yuck. If you haven't checked earlier, you need to check before uncapping and return these frames to the bees. Don't harvest brood. Period. You will also find some frames that contain pollen-filled cells, some with bee bread, and some empty cells. You will end up with enough pollen in your finished product without looking for more. Leave the pollen in the hive with the brood.

The easy solution to all of these problems is simply not to harvest these frames. Leave them in the field to be dealt with by the bees. Unripe honey, brood, pollen ... all diminish the quality of the finished product—pure, natural honey.

A frame with both honey and brood should be left on the hive. Allowing brood juice in your honey is simply a crime.

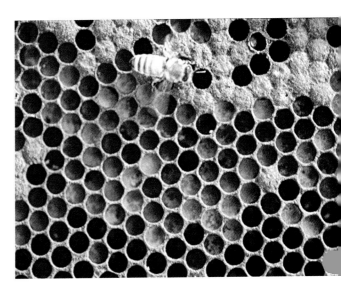

When you find pollen in a frame with honey, leave the frame in the hive. Honey has enough pollen in it anyway, and adding more means more to filter out, and more cloudiness in the finished product.

Chapter 4

Fragile: Handle with Care

THERE ISN'T MUCH HARM you can do to a honey crop while it is still in possession by the bees. Oh, you could feed sugar syrup at the wrong time so it gets into the honey. Or you could medicate your bees at the wrong time or with the wrong medication and some of it may contaminate your crop. You could even use too much smoke when examining honey supers or harvesting the crop, and that could taint the flavor of the honey even before it is harvested. But that's about it; that's about all you can do to harm this delicacy. You have to try hard to harm honey on the hive, and you have to want to do it.

THE DARK SIDE

Rarely, your bees can produce a honey crop that is less than ideal. They may visit plants that produce an off-flavoured honey—the northern variety of privet, for instance, produces a honey that is favoured by the bees but has a very disagreeable flavour for humans. There are other crops as well that produce unpleasant-tasting honeys that you should be aware of: almonds in California and other parts of the world, Brazilian pepper in the southeastern United States, and tarweed in the desert southwest of the United States are some that come to mind. There are others, certainly, and they should be avoided. Some varieties of honey are favoured locally but strongly disliked fifty miles (80 km) away. If you alone are consuming your crop, then this isn't a problem. However, if you are selling your honey, identifying both the nectar source and where it was grown is strongly recommended. Selling a bottle of strong, bitter, mixed aster/goldenrod honey to people who expect the variety of goldenrod that is more like butterscotch is a disservice to them and earns you a reputation not fully deserved.

Further, there are some plants that will produce honeys that are actually dangerous to humans. The classic example is the Kalmia rhododendron that grows in the forests of the eastern parts of the United States. Generally, honey bees do not favour these plants because there are so many other flowers blooming at the same

IT'S A MATTER OF TASTE

Just because the bees seek a particular honey source does not mean it is the best of all honeys. In fact, it may yield a terrible tasting honey. But honey tastes can definitely be regional. You may find a particular honey totally inedible that is a favorite of people only one hundred miles (161 km) away. Moreover, what most consider universal honeys, such as clover, can have different flavors in different regions due to climate, soil type, farming practices, and clover varieties. Local honey usually sells well locally, but be careful when you venture past your horizon. The truth is in the taste.

The first step when producing handmade honey is to ensure your bees are healthy and productive. The curse of varroa mites, various diseases, and pests can all play havoc with the productivity of your colonies. Is this a good queen? Will this colony thrive? If you don't know what's wrong, revisit the basics so you do.

time. However, on rare occasions, after a very late frost, for instance, many of these other blossoms will have been killed, and all that is left are Kalmia blossoms. These flowers produce a poisonous alkaloid that can harm and even, in extreme cases, kill a human. There are other flowers that have the same effect, so you need to learn of these, too, and avoid them at all costs.

But even all three of these combined—contamination, off-flavoured, or even poisonous honey—add up to almost no problems ever. Basically, it is difficult to harm honey while it is still in the hive. The bees have designed it that way.

But it's after we harvest it that honey's exposure to danger begins. We have figured out an incredible array of ways to make a perfect crop less so, and in far too many cases, we can absolutely ruin it. There is a golden rule in beekeeping: as with the practice of medicine and doctors, First, Do No Harm.

Keep It Cool

The entire focus of this book is how to harvest a honey crop that is as good as what the bees have produced. To accomplish this, we must know the things that do harm to honey, and the greatest harm we do is to heat it past its optimum temperature, which is no more than 100°F (38°C). Less is better certainly, but the

Your honey should be routinely examined with a refractometer for moisture content before it is harvested, during the uncapping and extraction process, and after it is extracted but before it is stored. Make sure your refractometer is calibrated before each use.

chemistry of honey is such that it well withstands that temperature because it is routinely exposed to an extreme such as that in many parts of the world. Bees themselves will cool a hive's interior to 100°F (38°C) or less when the ambient outside air is warmer so they keep their crop at its best.

Once harvested, the opportunity for contamination with sugars is nonexistent unless the crime is deliberately committed by the beekeeper. Water, sugar syrup, corn syrup ... no other sweeteners have a place mixed in with a honey crop. Nor do pesticides used to control those creatures that prey upon honey supers, combs, or even the honey after it has been harvested have any place in this process. Prudent sanitary conditions, timely extraction, and mechanical restrictions should be enough to keep these pests at bay until the crop is harvested and safely stored in bottles, pails, tanks, or drums.

Keep It Dry

Because the honey we harvest must be the best there is, the moisture content must be monitored continuously to ensure the honey will not ferment. Start checking in the field and continue until you have it bottled for sale. Guessing is not good enough. Seemingly capped honey can indeed have too much moisture. Bottled honey can ferment and even explode. Measuring is the only way to make sure.

And finally, how we treat honey that has crystallized is a measure of our skill as a beekeeper and as a honey protector. We can, if we wish, hurry the warming process so that it goes from solid to liquid in record time. In the process we will without fail heat it too hot, too fast, and too much and damage the honey. Rather, a gentle, slow, and easy warming process should be employed so that we don't drive off those aromatic compounds, don't degrade the enzymes, and don't change the colour.

All of these are ways we can harm honey. There are no excuses when it comes to causing harm to the crop we and the bees work hard to produce. We simply need to be aware of the physical factors that can occur and what chemical changes will result from those forces so we do not damage or destroy our crop. Remember, your role as a beekeeper once you have harvested your crop is First, Do No Harm.

HANDLING EACH CROP

Beekeepers who produce only an end-of-season honey crop—that is, they harvest and extract only once, removing all of the season's accumulated product from their hives at the end of the season—miss out on an incredible variety of honeys they could have harvested. A one-time harvest tends to blend all of the honeys the bees have gathered—the light and mild early spring bouquets, the midsummer wildflower mix, and the hearty late-season crop—and robs each of its identity. Moreover, it masks the unique flavours and aromas of the pure varietals that could have been.

The goal so far has been to suggest that honey should not be harvested as a single, blended, undistinguished medium-coloured, medium-flavoured crop. The techniques, timing, and biology shared allow you to home in on certain crops during the honey season. This way, you can produce early, mid-, and late crops of excellent honeys, even if harvesting true varietals isn't possible. Of course, many places have only one or two short, intense honey flows, and it's difficult to separate them. But even so, you should be able to pinpoint some varietal production schedules so those very unique and special crops, too, are available to harvest.

By now you know that if you are producing honey to sell, especially to a discerning clientele, the more types of honey you have available, the more honey you will be able sell.

Making Room for Variety

Having several different crops of honey available offers a world of increased sales opportunities. More often than not, a store or market will make room for each variety you have—rows for light, medium, and dark, especially if they have a seasonal or varietal name on the label. And that name is important; it gives repeat customers a goal—"I want that Spring Blossom kind," rather than just, "I want the kind that was here last time."

So producing a seasonal or varietal honey has distinct advantages when it comes to selling honey, and if that is why you are keeping bees, then you can see the value in paying attention to the tricks of the trade explained so far. However, producing these different kinds of honey has other advantages, and I hope you find value in being able to enjoy a light, mild early spring variety in one of your favorite recipes, or being able to offer a rare variety enjoyed by special friends or honey connoisseurs.

Caution: Overheating Honey

We routinely expose honey to temperatures that are simply sinful. A brief list:

- Hot rooms set so that the wax nearly melts (140°F [60°C])

- Uncapping knives set at boiling temperature

- Flash heaters set at 150°F (66°C) or more

- Harvesting brand melter honey to sell

- Heating bottle line honey to more than 100°F (38°C) for ease of bottling

- Heating crystallized honey to more than 150°F (66°C) for hours to hurry the process

- Using band heaters on pails or drums

- Storing bulk honey in barrels or pails outside in the summer, in the sun

But sometimes this just doesn't happen. Sometimes, some years, some places, being able to separate early, mid-, and late season crops doesn't work. The weather runs them all together, your schedule doesn't even come close to the schedule the bees are on, your equipment breaks down, your help takes a vacation just when you need them the most ... simply, life gets in the way and it doesn't work.

You can, of course, promote your crop as a once-in-a-lifetime blend of honeys that will never occur in just these proportions again, and that if customers find this year's crop enjoyable, they should consider buying more now, because there won't be any later.

Labeling Supers

There are other ways to take advantage of a one-time harvest, but it takes a little extra work when you begin processing your crop. In fact, it actually begins in the field, before you remove the honey supers from the hives.

If you simply remove supers—no matter the technique you use to coax the bees from the frames—and stack them in your truck or in the trunk of your car or just carry them home in a cart, you will lose their time of origin, that is, the exact time when they were filled by the bees. To reduce the confusion at harvest, label each super; use a wax pencil, a magic marker, chalk—whatever works and will last long enough to get from the field to extraction. Number each super beginning with the first super you put on each colony at the beginning of the flow. Try a large number "1" on the two narrow sides of the first-applied honey super (because that's how they enter the extraction area, narrow end first). The super above it is labeled "2," and so on. That way, the position on the colony of each super that comes home is noted.

When the supers are in the warming room of the honey house, group all number "1" supers together, the number "2" supers, and so on. When ready to

HIVE HINTS Selling Handmade Honey

When selling honey at a farmers' market, county fair booth, or anywhere you have one-to-one contact with a customer, try this technique. Obviously you should always offer a free taste test. "Would you like to try this light, mild variety before you buy?" is your question. But immediately follow it up with, "But you know, I kind of like this somewhat flavourful summer bouquet, too. Would you like to try this one, too?" And offer a second variety to try. Then, while they are still deciding (and they may want to try the first one again), ask this simple question: "Which one do you want?" Not, "Which one do you like best?"

Water White	0 to 8 mm
Extra White	9 to 17 mm
White	18 to 34 mm
Extra Light Amber	35 to 50 mm
Light Amber	51 to 85 mm
Amber	86 mm to 114 mm
Dark Amber	greater than 114 mm

Although honey exists in an infinite variety of colours depending on the floral source, treatment during harvesting, and age, these colours are graded. These measurements standardize the colour when honey is sold so both the seller and buyer know what they are getting. They are measured on a Pfund grader, a device that gives a reading of the colour in millimeters. (Extra light amber is the most abundant colour of honey and is often referred to simply as ELA.)

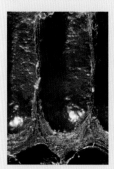

The comb that honey bees store their honey in can influence the colour of the honey harvested from that comb. Old comb, rife with propolis, soiled from travel, perhaps once used for brood so it is filled with cocoons and frass, generally turns dark brown to black after several seasons. Much of this material is water soluble, so when honey is stored in these old combs, the soluble fractions will dissolve in the honey and darken the colour. Combs in honey supers should be replaced every three or four years to avoid this darkening problem.

begin, move the number "1" supers out to the uncapping area first and break the frames free from the boxes. Then, pull frames and hold them up to a light. Separate the frames, if any need separation. There are usually only light- and medium-coloured frames to deal with, but a dark few may sneak in, so there may be three groups. This way different colours, and generally different flavours, are separated. The number "1" supers tend to produce honeys that are generally alike when separated into different colours.

Of course, you need to check the next numbered supers from the same colonies to see whether there is overlap, and often there is. The best way to tell for certain whether a medium colour from a number "1" group is the same as or different from a medium colour from the number "2" from that colony is simply to taste them. Do they taste the same? If so, put them together. If not, make another pile.

These tips are offered early in this section so you arrange your honey house before you begin separating frames and supers when you are stacking, uncapping, and extracting, rather than after you begin and there's no place to turn around, let alone make two, three, or more stacks of mostly alike honey.

PREPARING THE HONEY HOUSE

There are some hard-and-fast rules to follow when setting up a honey house. Many beekeepers do not have a dedicated location where they keep their honey-processing equipment year-round. It is convenient beyond measure if you have that luxury, but even so, these rules still apply.

Before you begin, you must be familiar with the health codes that apply to your operation. These may be city, county, or some other government jurisdiction. Your facility may fall under several bodies of governance—the department of health and the department of agriculture, for instance. These rules and regulations may overlap, or each may miss certain points; it all depends on where you live and who is in control. They vary from incredibly harsh and strict for even a two-colony hobby beekeeper to a live-and-let-live attitude for even the biggest operation. There is no rhyme or reason to how different localities handle the same situations, and there is almost never any negotiable latitude in how they are enforced. You must know and follow the rules.

Moving hives from honey crop to honey crop is easy with a portable setup such as this. There is storage room within the trailer for extra equipment, and at honey harvest the hives do not need to be loaded or unloaded. Mobile beeyards are more common in Europe than in the United States.

When Unloading Supers

Doors going into an unloading area, and all doors inside between the rooms, should be wide enough to easily accommodate the widest side of a 10-frame super with several inches to spare, or wider than the hand cart you are using, for easy handling.

You should provide more than enough room to move supers full of honey to the uncapping area and still have room there to move around and move frames out of the supers. Plus, you'll need room nearby to stack supers once they are empty. Lots of room is the rule, or lots of handling is the result.

Often, however, honey production falls under the jurisdiction of reduced regulation and surveillance by the food police. This is common with honey, maple syrup, and sorghum in the United States, and many other nonperishable crops. But don't bet on that always being the case.

The size of your operation will often place you in a less-governed situation, too, because inspectors are inclined to favor stricter oversight of larger operations where a greater number of people will be affected if a problem should arise. They have only a finite amount of time and would prefer to spend it as efficiently as possible. Moreover, honey is a product that simply doesn't cause many large-scale problems, period. It doesn't spoil or contain harmful by-products, and for the very few times it is found to be contaminated, the contaminant is almost always a dilutent such as corn syrup, sugar syrup, or water, and thus harmless (though illegal). The record of honey causing health problems in humans is essentially nonexistent. As a result, it seldom rises above the radar for health officials. But again, don't count on never being examined, especially if your product enters mainstream commerce such as a grocery store or farm market.

A set of guidelines has been developed by the honey-processing industry that you should be aware of. They will eventually be adapted in some form for honey processors everywhere. Below is a brief review of these. It is important as your business gains size not only to be aware of these, but also to check and update your facility to conform as these rules change. Being ahead of the curve is far better than being chastised by local officials because you weren't paying attention.

Industry Guidelines

- As a rule, check your building's exterior for problems leading to contamination (contamination here is simply extraneous material coming in contact with your honey, rather than deliberate contamination by the beekeeper). Natural and safety-type artificial lighting should be adequate, along with excellent ventilation.

- Sewage and water drainage systems should be up to code, along with employee (you) washing and sanitary facilities. And bees should be kept outside and away from the product.

- Hot and cold water should be available, along with adequate floor drainage and sink disposal.

- All nonfood chemicals should be stored safely away, and honey supers should be stored separated from the food-handling portion of the building. All equipment should be such that it can be cleaned.

- Supers and frames should be protected from pests and environmental contaminants such as insects and dust, and everybody who works in the facility should be trained in how to operate the equipment properly for safety and cleanliness.

- If your product enters the mainstream of commerce, it should be identified such that a recall is possible. Scary, but necessary.

No matter what regulations you are working under, or even if you will never enter mainstream commerce, these basics should be a part of your everyday behavior. Your honey-processing facility must be clean and it must be cleanable. Hot running water is a must. Stainless sinks—at least two and now usually three, depending on your local regulations—should be available for washing and rinsing equipment, and walls, floors, doors, and ceilings should be made from material that can be washed down. Ceiling lights should be safety types, and doors and windows should be screened so that insects—honey bees—cannot enter.

THE FLOW OF THE HONEY HOUSE

A honey house design should accommodate an easy-to-implement flow of supers from outside to hot room to uncapping to extracting to unloading the extractor to removing the now-empty supers and handling the now-full honey containers or tanks. Doors should be wide enough to easily handle the wide side of a ten-frame super, and the floor shouldn't have ridges, bumps, or weather strips that will hold up a two-wheeled cart. Few things in life are more frustrating than hitting a bump with a cart full of empty supers and watching them spill all over creation in front of you.

There should be adequate room to stack full supers close to the uncapping station once they have been moved from the hot room. There should be a work surface to drop the supers onto that includes a protruding frame remover, a place to clean empty supers of prop-

You need a place to use your scratching fork. It should be solid, have a way to catch wax and honey, and be high enough that you can still work comfortably.

Once frames are uncapped, they will drip honey. Some sort of tank or catch should be available. Many uncappers have one attached or as part of the unit. This unit is available from Maxant and Betterbee (see Resources, page 164). Frames are lowered between spinning chains, and cappings are removed and fall below. Other models have a small wax spinner in place of a simple catch, which is ideal for saving wax and honey. This model has a conveniently small footprint.

A level loading area makes moving supers into the hot room easy and safe.

Free-wheeled conveyers are ideal for unloading supers from the back of a truck to a stack of supers inside, or to a surface easily reached by the conveyer. They make for fast unloading and little lifting. If you are working alone, stick a broom handle in the far end to keep supers from rolling off. Use plywood sheets under the supers so they roll easily on the conveyer.

olis and burr comb once they have been emptied, and a catch for the material you remove.

However you uncap your frames, the cappings should fall into a catch so that the wax and honey are contained and not underfoot. Most uncapping devices have this built in, but some don't and one should be provided. Moreover, if the catch is out of sight, there should be some sort of overfill warning included so that when the catch is full of honey and wax, it doesn't spill over onto the floor. Or the container should continuously drain into a sump tank or drainage line to separate the floating wax and the honey underneath.

Once frames are uncapped, there should be a place where top bars can be scraped clean of propolis and burr comb. This residue should fall into a catch tank or trap so it isn't underfoot and any liquid honey is held so it doesn't run on the floor. These frames are already uncapped, and while waiting to be placed in the extractor, they will be bleeding honey. That honey should be captured and returned to the storage system rather than mixed with the wax and sent to the melter.

Often frames have capped cells that are difficult or impossible to remove with a knife or machine, and they need the attention of a cappings scratcher. This often occurs in the same location where frames are cleaned after uncapping but before entering the extractor, so the same catch can be used for the residue from both the frame cleaning and the cappings scratcher.

There should be a place to stack empty supers near the extractor so the now-cleaned and extracted frames can be placed back in the supers and then the filled supers can easily be removed from the room.

Once empty supers are refilled with empty frames, they need to be moved out of the way so others can move to their place. This should be a seamless move with a cart full of empty supers leaving the room immediately followed by a cart with full supers coming in.

Supers that contain empty frames should be removed to a location outside or where water on the floor and wax in the drain aren't a problem, so they can be washed and separated into stacks needing repair, paint, or no further attention at all. A power washer or hose with a direct spray nozzle should wash down each super so all honey and wax is removed from the outside of the supers and any loose paint is removed at the same time. You are lining up your winter chores here, but this makes it easy to identify the scope of the tasks needing to be done later.

Even if you are dealing with only a few hives, the concept of an organised and efficient extracting system should be in place. Your work will be so much easier and will go so much faster that you will soon decide you can handle more colonies, more supers, more frames, and more honey. And indeed you can.

Let's examine each of these steps in more detail so you can see how they work. And let's take a look at a good sample of the equipment you can purchase to do all of these things.

The Unloading Area

Dealing with your honey house begins before you get inside. What sort of unloading area do you have? Leaving supers stacked in the back of the truck in the driveway doesn't work because soon you will have every honey bee in the county examining them for any nook or cranny or spilled drop of honey they can rob and take home. If they find some, and they probably will, they will be back—with friends—and soon you will have a melee in your driveway that you, your neighbors, and your family do not want to contend with. So as soon as you arrive, you need to unload the supers and get them inside.

Carrying the supers from here to there is the least expensive, and hardest, to do. A simple two-wheeled cart is fast, inexpensive, and labor intensive. Specialized two-wheelers are available that fit a single colony pallet/drip board—forklift style—and work well for this. Motorized two-wheeled carts work better if you have some distance to move supers, and you can find or make one. Moving up, there are pallet movers and forklifts, if your operation warrants this scale of equipment.

A loading dock arrangement is ideal, where the floor of the honey house is level with the back of your truck, so you simply unload on one level. The unloading dock may be constructed below grade, so it is on an incline, or the honey house floor may be a raise above grade.

Some use manual conveyers to move supers from the back of the truck to inside. These are commonly used in grocery stores and warehouses; when available, used models can usually be picked up for a song. They save an awful lot of lifting.

Once supers have been emptied at the uncapping end of the line, they should be moved to the extracting end of the line so empty frames can be replaced and the now-empty supers moved to cleaning.

After empty supers have been removed from the extracting area, they should be cleaned of all honey residue. Washing with a power washer or spray hose will clean supers and remove loose paint. Supers can then be stacked to show supers needing no repair, those needing paint, those needing other repairs, and those that should be discarded completely.

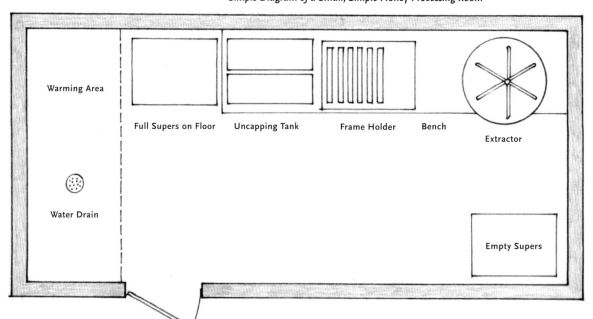

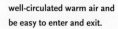

Simple Diagram of a Small, Simple Honey-Processing Room

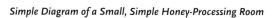

Warming Area

Full Supers on Floor   Uncapping Tank   Frame Holder   Bench

Extractor

Water Drain

Empty Supers

When efficiently designed, even a very small space can be used to process honey. The full supers are moved into the area on the left, which used to be a shower stall. A curtain holds in the heat from a space heater. When warm, the supers are moved out and stacked near the uncapping tank. Uncapped frames are moved to the frame holder and cleaned while waiting to be extracted. Extracted frames are moved from the extractor to the now-empty supers, and the extractor is filled with uncapped frames. Full supers are moved out with a two-wheel cart, and more are moved in to warm.

One option to consider, perhaps in the future, is constructing your honey house so that you can drive or back your truck inside to unload. You might set up an unloading dock inside, use a conveyer, work with a cart, or simply carry them to unload. This solves a lot of problems relative to attracting bees to your location and being able to leave a loaded truck overnight in a location protected from the elements.

The Hot Room

Once inside, supers to be stored for any length of time—no more than two or three days if small hive beetles are a problem—should go into a hot room, or makeshift hot room. This can be the simplest of spaces, such as a small shower stall with some means of enclosing the space and a small heater included, to a gymnasium-size room complete with built-in heaters, fans, and dehumidifiers. Most of us have something in between, such as a corner of the garage, or just inside the door of the honey house. Recall the earlier discussion on separating supers, so be sure to leave room, or make room for this activity if required.

If you are able to set up a hot room, the goal is to make and keep the ambient temperature warm so extraction is simplified. If possible, increase the ambient temperature to 80°F to 100°F (27°C to 38°C) and at the same time reduce the relative humidity to as low as possible with moving air or a dehumidifier running for twenty-four hours or more. Warm honey flows faster and better than cold honey does, and the drier air will reduce small hive beetle problems and lower the moisture of any too-wet frames. When you are ready you want your honey to be ready—warm, dry, and bee-free. If you harvest later in the season when outside temperatures turn cold, your honey may need two or more days to warm to the desired temperature.

However, if you don't have a setup that enables you to warm the honey, process it as soon as you can because it will still retain some of the hive heat it had. If this isn't possible, then a small space heater in the vicinity of the stack of supers could be used, but only when there are people around to monitor it; don't leave it unattended. You may be limited to ambient temperature, such as an unheated garage. If you know this ahead of time, then you can plan on processing as soon as possible, or plan on a much slower process if the honey reaches ambient air temperature and it's cold outside.

UNCAPPING

Before we look at the specifics of this task and the equipment choices you have, you must appreciate the effect this equipment can have on the final product you are producing: handmade honey. Here are some things to watch for.

Cappings Scratcher

There are several models on the market that all basically do the same thing. They vary a bit in the number of tines, and more in the style of the handle. Caution: the tines are very, very sharp.

- What is the temperature of the hot knives? Make sure they don't scorch even a tiny bit of the honey beneath the cappings you are removing.

- How much cappings wax will end up in the honey mix when you are finished extracting? This is a function of the removal process and can negatively affect the final product.

- How well are the cappings being removed? Too shallow, and you'll need to do more scratching, adding wax to the mix. But if the cuts are too deep, you may damage frames (adding wooden splinters to your final mix), but certainly additional wax will fall below.

- If you are using automated equipment, can the speed and width be adjusted to accommodate different loading speeds, different frame sizes or styles, and different comb thicknesses?

- Check the uncapper's footprint. Is there enough space to work around it—one side, front and back? And how accessible is the container holding the removed capping and honey? Is the model you are considering too large for your space? Can it be easily moved when not in use? Will you have to upgrade, increase, or otherwise change electrical or drainage attachments?

- What about potential growth? Will this machine accommodate a larger or faster extractor down the road?

Uncapping equipment runs the gamut from a simple table fork to a special uncapping fork to serrated bread knives to unheated off-set uncapping knives to heated offset uncapping knives to vertical or horizontal flail uncappers to unheated hand-operated or mechanical punch and vibrating knife machines to heated vibrating knives to whirling blades or rows of spindles on spinning shafts.

Coupled with this equipment are the containers that capture the beeswax cappings and honey that are removed in the uncapping process. These containers can be as simple as a plastic 5-gallon (19 L) pail, plastic tubs, double tubs with strainers and honey gates, stainless steel tanks on legs with strainers, or extended tanks that automatically move any number of frames toward the extractor. Both an uncapper and a catch tank (separately or as a unit) can fit most any space, budget, and size of operation.

Removing Bees From the Honey House

It's almost impossible to bring supers of honey from a beeyard to a honey house without some adult bees joining you. A persistent few always manage to evade blowers, overcome fumes, avoid brushes, and figure out escapes. This is especially true if there is any brood in even one of the supers.

Flying bees inside a building are a nuisance, and if you have inexperienced helpers, these loose bees will be distracting and slow down the work. Don't jeopardize the well-being (real or imagined) of your help—give the bees a place to go. The best choice is a bee escape mounted in one section of a multisection window. But if you harvested honey several miles away from the honey house, bees that escape have nowhere to go and will simply hang around the outside, attracted to the smell of food inside. Or they'll rummage around the neighbourhood, looking for trouble.

If installing an escape isn't possible, bees that gather at a window can be dispatched using a soapy water spray, or flypaper hangers, or with an electric bug zapper. Any of these methods work, but they unnecessarily kill honey bees. Another thought is to provide a trap hive for those bees. This can be made so bees inside enter from a one-way escape into the hive, perhaps through a hole cut in the wall with an escape installed, and there is free access from the outside so that you are capturing wayward bees from both locations. These bees can be used to boost weak colonies, or with the addition of a frame of brood and a queen, start a nuc or small hive.

A final choice is that you can accept that these wayward bees are part of the uncapping and extracting equation and will be in your honey, in your hair, and on the handles of pails and doorknobs. This should make the removal decision easy. However, in no case is the use of an insect spray acceptable. No sprays, ever, in a honey house.

In any event, live bees in the honey house tend to become dead bees in the honey house, and dead bees end up in places you don't want them.

Bees inside the honey house are attracted to the light outside coming in from the screened area through a one-way escape. Once outside, they are no longer a problem inside, and are attracted to the hive placed conveniently in the upper corner. When full of bees, the hive is replaced, keeping the bees out of everybody's way.

Unheated offset serrated uncapping knife

Unheated offset nonserrated uncapping knife

Uncapping frames and capturing the wax and honey that is removed should mesh well with the capacity of the extractor being used. A slow or inefficient uncapping station paired with a fast or efficient extracting operation, or vice versa, makes for inefficient use of your time. Add to this that sometimes you are working alone and sometimes you have help. This will change the metric of the entire operation—uncapping, loading, and unloading the machines you are using and your overall efficiency.

Another factor is the temperature of the honey you have. The value of a hot room becomes obvious if you ever have to handle honey that's cold to the touch—it just doesn't want to move very fast.

If you already have the uncapper of your dreams, then this will serve only as a review. But if your uncapper is underperforming for your current situation or you are in an expansion mode and moving up, it will pay to take a look at these basic styles of uncappers. I'll briefly point out the advantages and disadvantages of each.

Hand-Powered Uncappers

The simplest uncapping tool is a kitchen fork. It is also the least efficient. A cappings scratcher is similar but certainly more efficient. There are several models that range in width, with straight plastic handles to curved plastic handles to easy-grasp wooden handles to cushioned rubber handles. Consider how much use it will receive and get the best one you can find. The old rule applies—you get what you pay for. One difference to consider is that smooth plastic handles will become slippery with honey after repeated use. Regardless, no matter what kind of uncapping setup you have, an uncapping fork is necessary. Remember, though, that one adds wax residue to the remaining honey in the frame that must be filtered out later.

A SERRATED BREAD KNIFE will work if you have only a very few frames to uncap, but it becomes impractical if you have a larger operation. The flexible blade is the biggest problem, and because the handle isn't offset from the blade, you'll end up with skinned knuckles and a cramped wrist after a short while. Also, the handles on these knives are not made to accommodate the necessary angle of cutting and soon become very uncomfortable, and even unsafe when they're slippery with honey.

AN OFFSET SERRATED KNIFE is better adapted to removing cappings because it is sharp, has a wide blade to discourage capping flowing over the blade and sticking to the cappings on the other side, does not flex, and eliminates, mostly, the bruised knuckle problem. But it, too, has limited use. If you choose this method, seriously consider getting two or more so you can be heating the blades of one or two in hot water while using the other. A hot knife cuts through the wax easier than a cold knife does. These knives are sharp. Grabbing one the wrong way, having one slip out of your hand, or dropping one on your foot can cause serious damage to exposed fingers, hands, or toes. Moreover, most serrated models are made for right-handed people. If you're hopelessly left-handed, don't even try one of these. And because they are serrated, they are difficult, perhaps impossible, to sharpen after a few seasons' use. Using a dull tool is dangerous. When dull, discard and replace.

AN OFFSET UNSERRATED COLD KNIFE (here, cold means there is not an electrical heating element in the blade) is more useful. When using this, it's also wise to have one or two knives heating in water while using another hot knife so the blade moves easily through the wax. They are heavier than serrated knives, so the weight of the tool gives you some additional leverage, and the blade isn't flexible at all, which also helps. The large handle is made to accommodate the angle at which you are moving the blade, and you can grip it, so it won't slip or turn while cutting through burr comb or thick wax. They are more durable than some types of knives you can use, but they, too, have their limits. Primarily, you can cut only so fast and your arm can last only so long before tiring, though it is amazing how many frames you can process in a day with one of these. They become dull with use but can be sharpened, and it is advised that you do so periodically. A sharp knife is always easier to use than one that is dull.

Heated Knives

THE ELECTRICALLY HEATED OFFSET UNCAPPING KNIVES are a step up in efficiency because they too have weight and width helping out, and because they're heated, they will cut through wax cappings more easily than unheated knives will. Heated knives are efficient, but they have drawbacks to be aware of.

Heated offset uncapping knife with the thermostat in the blade

Heated offset uncapping knife with an inline thermostat, the top-of-the-line model

UNCAPPING TRIVIA

....

- *Some books call this step 'decapping', though most sources use 'uncapping'.*

- *When uncapping with a knife, holding and moving the knife from the bottom of the frame toward the top of the frame at about a 30-degree angle tipped away from you works most efficiently. As the wax cappings are released by the knife, they swing away from the frame and fall into the catch.*

- *If you fill a super to its maximum capacity, (eight in an eight-frame super), the bees will 'draw out' the beeswax comb so just enough space for one bee remains between the surfaces of adjacent combs. As a result, the comb that extends past the edge of the frame is very short, leaving very little for an uncapping knife (or blade) to remove. Under filling a super (six or even eight frames in a ten-frame honey super) will encourage the bees to draw out the beeswax cells more, filling more space between combs. This makes the removal process much easier and leaves few, if any, sunken spots on the frame's surface.*

- *Always have a pail of water nearby when uncapping. Honey is sure to drip making everything both sticky and slippery, and dangerous. It will cover your hands; the floor; the edge of the cappings catch; and more. A pail with water and a washcloth nearby makes continuous cleanup easy.*

- *The handling principles for beeswax apply to cappings, too: never heat wax and honey together, and don't overheat the wax, as it will darken. Always clean up immediately so the unattended cappings do not attract wax moth and small hive beetles. Always remove as much of the honey as possible.*

- *Some beekeepers clean their cappings after draining by giving them back to the bees who will remove all of the remaining honey. Place these cappings inside a colony, usually on the inner cover, rather than outside the colony. Free access can trigger robbing.*

Generally, the heat settings are troublesome. Some models have one heat setting, and it is automatically set at too hot. More expensive models have crude thermostats set right on the blade that run from cold to not hot enough to too hot to way too hot. In short, they are hard to calibrate. Of course, too hot means that as the knife glides under the cappings, the honey is overheated. One common model even boils the honey it touches. Better models have sensitive temperature controls built into the wiring so the honey isn't compromised. However, all models have the problem of handling honey left on the blade when the knife isn't being used. The unit must be set to cool or cleaned off after every time it is set down. Otherwise, the honey on the blade will caramelize and then burn, and while this isn't debilitating, the burned honey on the blade will affect the honey in the frame when next used. Plus, the burned honey slows the knife running through the cappings on subsequent frames because it is rough. And just try and clean burned, caramelized, petrified honey off the blade.

There is another type of uncapping device commonly called a honey punch. Basically, it works like a paint roller, but rather than having an absorbent pad as a roller, it has a hard plastic base with protruding pegs. The roller glides over the face of a frame, and the pegs punch through the cappings as they pass. Rollers come in common frame widths, so no matter which frame size you use, there is a roller punch that fits so you have to go over each side only once (well, usually twice, once up and once down to get all the cells). There is essentially no cappings wax to deal with when using these, and the honey doesn't leak from perforated cells.

Wax Cappings and Honey

When you are uncapping frames of honey, the wax cappings and adhering honey must somehow be captured. There are several uncapping tanks on the market that allow (most of) the honey to drain from the cappings wax by gravity, hold a good deal of material so emptying is not a constant chore, and have a strainer situated in the container to hold the wax on one side so the honey can run out and a drain in the bottom to let the accumulated honey drain away. The differences are primarily where the unit sits and how large it is. Is it meant to sit on a workbench surface, does it roll, or is it fastened to the floor? And how many uncapped frames awaiting extraction are they able to hold? Some plastic models have several uses, and some stainless steel models double for other purposes as well, so you get your money's worth.

Your choices when using hand-powered uncapping technology are quite varied, but your greatest considerations will be where you can place the unit when you are processing honey, the footprint it will have either in the room or on the workbench, what other uses it has that are beneficial to your operation, and certainly not last, the cost of each model. You also have to take into consideration where it will be when it's not being used—how much space will it take up, and where will that space be? However, because none of these units will harm honey, they should all be considered.

Hand-Powered Mechanical Uncappers

These are uncommon but interesting machines. A couple of uncapping machines on the market are hand-powered but mechanical in nature. One is horizontal—that is, the frames are laid on a moving belt powered by a hand-turned crank. The belt carries the frames so they pass between two moving cylinders, one above the frame and one below the frame, that have pegs that puncture the cappings as the frame passes between them. The cylinders are mounted such that some movement is allowed to accommodate frames of varying thicknesses, and there is a small catch basin below the belt to capture any honey that leaks out during the process. Because the cappings are punctured rather than shaved or flailed off, very little wax enters the process, yet the honey easily escapes the cells when extracted.

The other model is vertically oriented and drops a frame between two serrated, moving knives, one on each side of the frame. Cranking the machine twice lowers the frame between the knives and moves the blades. This is a Cowen-based machine that operates much like the rest of their uncappers, with the exception that it is hand-powered rather than electrical. The guides that keep the frame straight in the process have little flexibility, so ill-shaped frames tend to slow the process, but the gearing and speed are adequate. You will need a catch basin for the wax and honey that are removed from the frames with this machine.

Either of these machines is more efficient at uncapping frames—faster, easier, and safer—than hot knives or other hand-powered techniques, but they all require labor.

This uncapper can handle as many frames a day as you can crank. It's faster and safer than using heated knives.

A Cowen Silver Queen model extractor showing the frame guide in front, the blades below, and the chain they move on. This model is sitting on a table that is the support for a drainage tank.

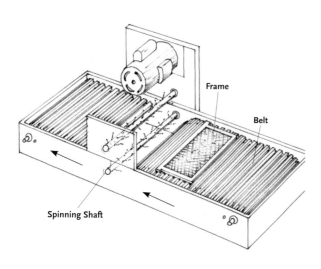

This cutaway drawing shows how the uncapper actually works. Frames are moved to the center on one belt. At the center, the belt on the right curves under. There is an empty space in the center where the spinning shafts are with their flailing chains. These chains whip the cappings off both sides at the same time. The moving frame continues to move on the belt on the left and is removed from the uncapper. The chain shafts can be adjusted for depth so there are no missed cappings, and the speed of the belts can be adjusted to accommodate higher volume or fewer people working.

Labels in drawing: Frame, Belt, Spinning Shaft

Automatic Uncappers

As a general rule, automatic uncappers are relatively safe when it comes to guarding honey quality, though if you try, you can do some minor heat damage with some of them. Basically, they come in two orientations—vertical and horizontal.

The vertical systems have a frame slide down an open guide until it is grabbed by a rotating chain. The frame is then dropped between two vibrating, heated blades that remove the cappings from each side of the frame at the same time. The frame, once free of the blades, is moved by the chain from the vertical drop to a horizontal position, so it hangs over a drip tank. Any loose wax or honey drips into the tank. Some models have the frames moved automatically after that, while others are hand moved toward the extractor. While moving or being moved, there is an opportunity to clean the top bars, scrape any low spots with a cappings scratcher, or remove broken frames.

Some models are bigger and faster than others, but the process is basically the same. The heated blades are, or should be, not very hot, and the heat serves only to keep honey and wax from building up while they are in use. Some beekeepers have modified the process and pipe a small amount of honey from the sump back to the blades to serve as a lubricant for the blade/wax motion. This further keeps wax from building up on the blade and speeds the process even more. Honey is not damaged in this process.

However, these uncappers tend to be hard on frames because they don't have a great deal of flexibility. Off-sized, irregular, or broken frames tend not to last long when run through one of these machines. And although irregular frames can jam and hold up the works, they tend more often to simply get eaten in the process. After a couple of seasons of this treatment, the consistency of the frames in an operation is remarkable. They either fit the uncapper, or they are kindling. Still, these machines can handle an incredible number of supers in an hour, and two people can be remarkably efficient at loading, uncapping, and cleaning an enormous number of frames in a day.

Another style of uncapper works both ways—that is, it can be either vertical or horizontal. This is the classic flail uncapper. For the horizontal style, a frame is laid flat on a moving belt. The belt moves the frame so it passes between two rapidly rotating cylinders,

Removing Wax Automatically

There are several machines on the market that collect the accumulated cappings and adhering honey when they fall from the uncapping device being used. Large-scale operations tend to use either of two methods to handle this sweet, sticky, waxy mess.

Capping Melter

One way to handle this wax and honey mix is to use a cappings melter that has a water jacket. Honey and wax fall into one side of a large, water-jacketed tank. The honey/wax mix is warmed, and melted wax is pumped off the honey, which remains behind. The temperature is such that it is hot enough to melt the wax (at least about 150 °F [66°C]). Honey heated this warm is certainly compromised, and even though this method is efficient for reclaiming the cappings wax, the wax can be harmed unless it is rapidly removed. These devices are designed for use in industrial settings and are geared more toward quantity than quality.

Wax Spinner

The other cappings processor is called a wax spinner. This device is much kinder to both the honey and the wax. When a wax spinner is used, the honey and wax captured from the uncapping machine are augured or pumped into a main carrying line that often has all the honey from the extractor(s) emptying into it, too. This honey and wax slurry then is carried to the wax spinner, which is simply a perforated cylindrical tank that fits inside a slightly larger cylindrical tank. It works exactly like a lettuce spinner. The honey/wax slurry is dumped into the inside tank, which is spinning rapidly. As the inside tank spins, the contents are flung against the inside wall of the inside tank. The honey oozes through the perforations, but the larger wax particles remain behind. The honey is captured in the outside tank body and drains away. The wax continues to spin until all the honey is removed. The wax ends up as a dry, coarse powder that is released out of the bottom of the tank and processed further. The unheated honey can then be further processed or stored. This is certainly a more humane way to treat a product as delicate as handmade honey.

A water-jacketed cappings melter

Removing dry beeswax from the bottom of a wax spinner. All the honey has been removed and processed safely without heat. A wax spinner sits at the far end of the uncapping and extracting process in this operation.

one above the frame and one below. Attached to the cylinders are short chains that strike both surfaces of the frame at the same time. The length of the chain is such that it just barely removes the cappings of the frame, opening the cells. This tiny bit of wax is captured below the belt and the frame moves on. One person can load and unload this uncapper, but it can be a task to keep up with because you can put two or three frames on at a time. There is no frame damage, only a little cappings wax generated, and not much dripping honey to deal with either. Only very recessed cells need attention from a scratcher. The frames must be moved to another location so they can be cleaned and stored while waiting for the extractor, so there is a second handling involved that doesn't exist with the previously described machines. But there is never a jammed machine, and there are no broken frames and dull blades to deal with.

Another uncapper, which has been manufactured for years, operates on basically the same principle but is, instead, vertical, so the frame moves down between the spinning cylinders with chains rather than lying flat and sliding between them. The wax cappings fall below. The frame moves down automatically or is raised and lowered by hand. The same advantages apply to this model. The only disadvantage is that the process is not continuous, and the unit can handle only one frame at a time. (See photo page 81, bottom)

There are other models available, mostly on the used market, that are similar to these. One has frames move between cylinders with pegs instead of chains, while others have blades on rotating cylinders, but the principal operation is the same for all of them.

Inside a Tangential Extractor

Frames are placed in the interior basket—this model holds three frames—so that one side of the frame faces the interior wall of the extractor. Hand-powered models are generously geared so even large, heavy loads are relatively easy to turn. Some, however, are easier than others. Honey that is spun out is flung against the interior wall of the extractor and runs down to the bottom, where it is collected. At the bottom is a gate from which the honey drains.

☞

Inside a Typical Radial Extractor

Frames are set in a regular radial extractor much more efficiently than in tangential extractors. Depending on the model, these extractors normally can hold fewer deep frames than mediums or shallows. Still, they are quite efficient. When the inside basket holding the frames is spun, the honey is forced out of the cells toward the outside of the extractor. There the honey is caught against the inside wall of the extractor and drains below. Because of the number of frames most radials can hold, arm power usually isn't enough and these units are generally motorized.

☞

EXTRACTING

The extraction process is probably the safest step there is in the journey of moving honey from bee to bottle.

There are, basically, two kinds of extractors: tangential and radial. They are easy to tell apart because they are constructed very differently. Radials are the largest and most efficient models.

Tangential Extractors

A tangential extractor has a mobile, spinning basket that sits inside a nonmoving cylindrical extractor. Frames fit in the basket such that one side of the frame is facing the inside wall of the extractor container. Generally these extractors hold from two to eight or so frames. To extract, the inside basket containing the frames is spun at a fairly rapid rate, either by manual power or by a powered unit. Centrifugal force spins the honey out of the cells on the side facing the inside wall of the extractor. About half of the honey from the first side of the frame is thrown out of the cells, is caught on the inside wall of the tank, and falls or drips to the bottom of the tank. The extractor is stopped, the frames are turned 180 degrees so the opposite side of the frame faces the extractor wall, and the basket is spun again. All of the honey from the second side is spun out. The frame is reversed again, and the remaining honey from the first side is spun out. You see the effort and inefficiency involved here.

The advantages of these units are that they are small, inexpensive, and easy to operate. They are generally a beginner's and small-scale operation's extractor simply because to handle the number of frames needed for even a moderate-size operation would require a tank as large as a football stadium. But that small size works in their favor in other ways, because they are relatively inexpensive, are easy to move and store, and can be quickly fastened down when using. Their size and simplicity makes them easy to use, even for children. Tangential extractors are a good choice for operations with perhaps as many as twenty colonies, but that would be pushing it for efficiency. For larger operations, one of the radial models would be better.

RADIAL EXTRACTOR CONFIGURATIONS

The photos show some of the ways various-size frames can be stacked into radial extractors. Some radial manufacturers can customize the frame holder to accommodate special situations.

As noted elsewhere, the size of your extractor, the number of frames it can hold, and the length of time it takes to extract the honey from those frames play key roles in the uncapping equipment that feeds the extractor, and the honey- and wax-handling equipment that the extractor feeds.

Equally important is how warm the honey to be extracted is, and how many people will be working. Idle time for equipment, or help, is wasted time and money. But don't forget the size of the unit, the location of electrical connections, and the plumbing required to host this entire network.

Plus, these units need to be both loaded and unloaded, and space will be needed for both full and empty supers, plus easy entrance and exit strategies.

A wise beekeeper once said, "For every hour you spend planning your honey house, you'll save five years of frustration and wasted time."

Many beekeepers standardize the size of the extracting frames they use and don't change them. If this is the case in your operation, purchasing a frame-holding wheel will be simple—get one that holds the maximum number of your size frames. But if you have two- or three-sized extracting frames, a wheel that can efficiently hold different sizes is what you need. Standard wheels generally hold all sizes, but some more efficiently than others. The top photo shows mediums, center shows deeps, and bottom shows shallows. Your operation will determine extractor size and accommodations.

☞

Radial Extractors

Think of a bicycle wheel when thinking of the two types of radial extractors. One type turns as if the wheel were lying flat, spinning on the hub. The other type turns as if the wheel were turning in an upright position, much like it does normally. For both models, the frames in the extractor are positioned so that the bottom bar is closest to the hub, and the top bar is closest to the outside of the wheel. This takes advantage of the fact that honey comb cells are angled upward about 17 degrees. The photos show it well.

Radial extractors are the workhorses of the honey industry. They can be small, handling six to ten frames or so, or they can be very, very large—handling 120 or more frames in a single load. The biggest units are industrial in scope and are so automated that just two or three people can handle hundreds of supers in a workday. One of the very largest has one person setting full supers on a platform that raises the frames out of the box, grasps them, and moves them to a long tank. There they are automatically spaced out so another person can clean top bars and scratch low spots on any frame that passes by. This tank holds 120 frames. When cleaned, they are bunched up and pushed into a parallel radial extractor. As one group of twenty frames moves in, they push out an already extracted group of twenty frames. These then are moved away, grouped in bunches of forty, and dropped, ten at a time, back into supers that have been fed into the system by the third person working on the machine.

You can see why it is important to consider the uncapper, the wax-and-honey slurry–handling equipment, any plumbing for moving liquid honey, any heaters and coolers, the extractor, sumps, filter or strainers, wax processors, and holding tanks as a single system. All the parts need to work in concert, or otherwise there are bottlenecks, backups, and stresses in the system. The large automated units come already powered and plumbed to handle the frames, the uncapper, the extractor, and the belts and drives that move the frames.

The plumbing required to handle the honey removed from this many frames must be substantial. Large, stainless steel piping, heavy-duty pumps and sumps, and high-volume inline heaters and strainers are required just to keep this much honey moving. The wax processor at the end of the line must be able to handle a significant amount of wax and honey unless a flail uncapper is used.

Another configuration for frames in a radial extractor. Models with small diameters can hold fewer deep frames, but many more medium or shallow frames. Knowing about this type of configuration before you purchase will help determine the depth of the honey supers you use, the speed of your uncapping operation, and the flow of honey from the extractor to the rest of the processing operation.

COMMON STYLES OF
RADIAL EXTRACTORS

....

Radial extractors are available in various sizes, speeds, and capabilities. They also come in various outside designs that may necessitate alternative honey house requirements. A one-piece construction with the legs attached directly may need a different setting than one that has an unattached stand. Diameter and height, too, need to be considered, and height off the floor for drainage or pumping to a sump should be considered before purchasing. Side drain and center drain will also make a difference in where pipes go and how much pipe you will need.

Nearly all radial extractors come equipped with a variable speed motor. Some have programmable speed drives so the spin starts slow, speeds up, then slows, all at predetermined speeds and for predetermined times. If these are available they are worth every penny they cost in saved time, work, and honey not extracted.

Some extractors, such as this small-scale model, are free-standing with attached legs.

Moderate-sized radial extractors, such as these nine-frame models, are small enough to work by hand, but certainly go faster if motor powered.

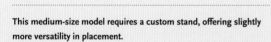

This medium-size model requires a custom stand, offering slightly more versatility in placement.

A sixty-frame extractor can handle more than one person can efficiently operate in a typical day when considering moving supers, uncapping and cleaning, loading and unloading the extractor, and moving empty supers out of the honey house. When purchasing a machine this size, your plan will call for additional labor and some level of automation to make the best use of its capability.

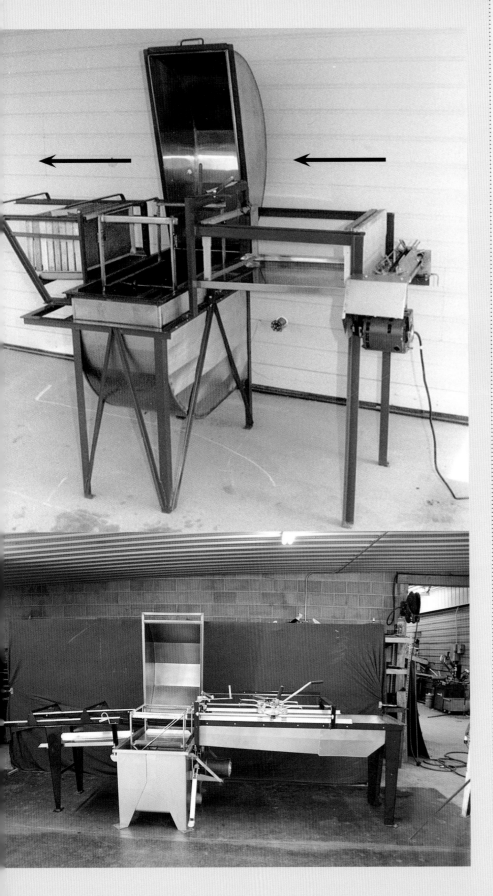

PARALLEL RADIAL EXTRACTORS

Parallel radial extractors are actually the most efficient extractors made. They become even more efficient the larger they get. Loading is done in bunches, as is unloading, rather than one frame at a time. Baskets are made to order so the extractor holds as many frames as possible for the size of the container. Plus, the footprint space/frame extracted is the smallest of any extractor.

Models are available that automatically move frames from an uncapping device over a drip and cleaning tank (for top bar cleaning and scratching) and then load the machine. When full, the frames are spun at predetemined speeds for a predetermined time—all programmed beforehand—and then unloaded in batches to be returned to empty supers.

If size, capacity, efficiency and space considerations are important, consider one of these. The three models illustrated on these pages represent the spectrum of sizes available.

The differences in the units shown lies in the number of frames each extractor holds. This determines the size of the frame feeding tray (shown left, above and right, opposite), and the frame receiving tray on the other side of the extractor. Size does make a difference in foot print, capacity, and cost. The arrows, show the flow of frames from beginning to end of the extraction process.

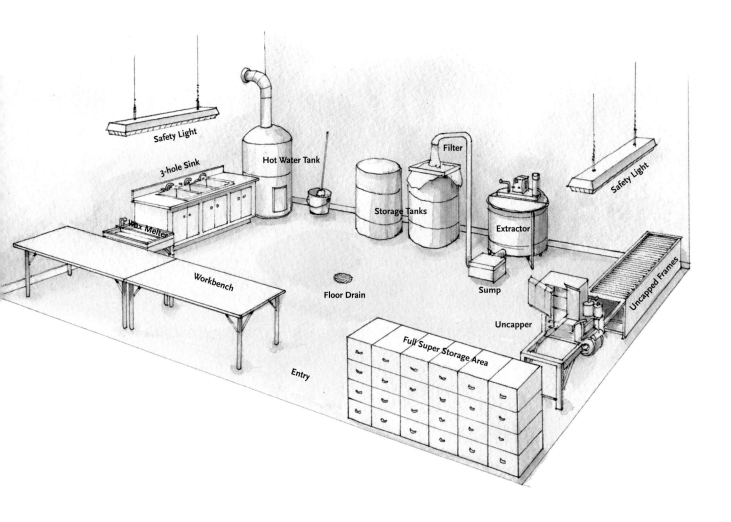

A typical honey house layout. Full supers are brought in from the warm room and stacked next to the uncapper, lower right. Once uncapped and cleaned, frames move to the extractor. Honey from the extractor flows into the sump, where wax is captured (and later moved to the melter against the far left wall), and honey pumped to a strainer and into a storage tank. A three-hole sink, water heater, workbench, and floor drain, along with a safety-shielded light, round out the facilities. There is ample room for empty honey supers near the extractor, so when finished, they do not need to be handled twice.

Troubles begin when each of these units is purchased separately. The uncapper comes from one supplier, the conveyer from another, the extractor from yet a third system, all hooked together with a homemade plumbing system—this is not uncommon in honey houses. Getting these to mesh, to fit well together and flow, can be a challenge, and before purchasing these individual pieces, you should give great care and consideration to each step in the process. Making an expensive mistake can be difficult to overcome in terms of time wasted, equipment broken, and frustration.

POST-EXTRACTION

Without a doubt, the worst crimes against honey are committed after the honey is extracted. Large operations are far more likely to commit felonious assaults on honey than are smaller operations simply because they must rapidly handle large volumes of honey. This requires large and extremely hot rooms, uncapping at the speed of light, extracting even faster, honey passing through the flash heater, then the filter, then the flash cooler on its way to the wax spinner. The heating, cooling, and heating cycle only diminish the quality of the product each time there is a temperature change, and by the time the honey gets to its final storage tank, it has been fairly abused.

Smaller operations, too, run the risk of getting honey too warm when it runs through a heated sump after emptying out of the extractor. A sump is simply a warmed (sometimes too-warmed) box that the honey slurry from the extractor flows or is pumped into on its way to being strained. Because the wax debris floats, it can be separated from the honey and is captured in a forward section of

Unheated Strainer

This model strains the wax from the extractor slurry without melting the wax or heating the honey. Wax stays behind, and honey moves on. This is treating honey the way it should be treated.

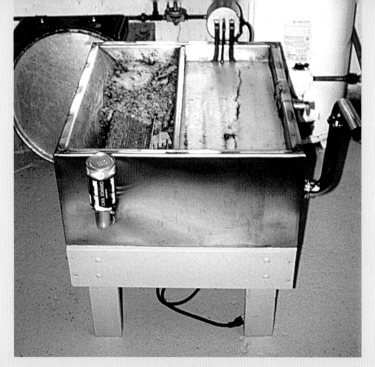

Heated Honey and Wax Melter

Generally, in smaller operations, the slurry from the uncapper is handled independently of the extractor honey, and that, too, can be damaged if it goes into a "brand"-type melter.

Some beekeepers capture the slurry from their uncappers in tanks or pails and set them aside for later attention. They may dump the entire solution on a strainer and let the honey drain away, then later melt the wax. Or they may simply heat the whole mix and let the wax melt and float to the top. Then they can remove the wax and dispose (we hope) of the honey, which has been far, far too overheated.

Simple strainers are fine material or metal mesh. For most small-scale operations, these work well. You can empty honey from your extractor directly into a 5-gallon (19 L) pail for storage or later bottling using the gate at the bottom. The cloth strainer keeps extraneous material from entering the pail. Most beekeepers will let the honey in the pail settle before bottling.

What one might call an "inline" strainer is supported by brackets just below the honey gate in the extractor. All honey passes through this strainer before being bottled or stored.

the sump's tank. The honey flows underneath and collects in the second section of the sump's tank and from there is pumped, (nearly) wax free, into storage or through a gravity filter. Either way, the wax remains behind and the honey flows away. Sump temperature control is sometimes variable, or, like the least sophisticated uncapping knife, has only one setting—too hot—and, ultimately, too-hot honey can result. Larger, more expensive models do separate the compartments, and the honey captured from these is seldom damaged.

Settling

There are several successful methods for removing those remaining bits of wax and residue from the finished honey that has not been ultra-filtered. The simplest is filtering to remove almost all of the bits. It must be explained here the difference between straining and filtering. Straining is a nonintrusive process where honey flows on its own accord through a strainer or strainers of diminishing mesh size. The honey may be as warm as 100°F (38°C), but no warmer, and it flows by gravity only. Usually there are a series of strainers the honey flows through, the first being rather coarse, no finer than window screen. A second strainer will be finer, usually constructed from a manmade fiber, such as a nylon or rayonlike material. There may be additional strainers of even finer material.

From there, the warm liquid is usually run directly into barrels or pails, where it is left to settle. Over time, any remaining wax particles or other particulate matter—bee parts, pieces of frames or propolis, or whatever else was small enough to make it through two or more strainers—will move to the surface of the honey in the container. Air bubbles that were incorporated into the honey during all of this commotion will rise to the surface also. Counterintuitively, this is known as settling, most likely because the honey itself is settling and all other material is rising to the surface.

Settling tanks are still used but are not as common as they once were due to improvements in both straining and filtering. Generally, settling tanks are shallow but have a large surface area. This speeds the process because any particulate material has only a short distance to rise. The same occurs in any tank or container if left long enough. A 5-gallon (19 L) pail works just fine, but it takes a while, depending on how warm the honey is when it fills the pail.

Strainers with very fine mesh, or several mesh sizes, remove more particulate matter, but still work for small-scale operations.

A double strainer. The first is a fine mesh nylon filter; the second is even finer and is attached to the wooden frame and supports the honey and the first filter.

Pressure filters remove most particulate matter, but the honey must be warmed to work well. These units can usually handle about 5 gallons (19 L) per minute.

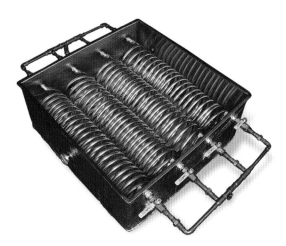

Honey is warmed with a flash heater by pumping it through pipes in a tank filled with temperature-controlled hot water or vegetable oil. From there, the honey passes through a filter, then on to a wax spinner or storage tank.

After some time, almost all of the material will have risen to the top, including the air bubbles incorporated into the honey during the extraction, pumping, or filtering process. This material appears as a film resting on the surface of the honey, sometimes crudely called scum. It is easy to simply skim off this residue using a large, flat spoon or spatula, or to simply drain the tank from the bottom, stopping when the film on the top finally gets to the very bottom. This allows you to get the best, cleanest, purest honey, and leaves the residue behind without further straining. Because of the mesh size of the strainers used, the fact that the honey is not heated during the process, and the fact that the tiniest of particles must rise to the top without further assistance, straining in no way affects the quality of the honey. The aromas, enzymes, sugars, some of the pollen, and all the rest remain unchanged and undamaged. This is how honey should be treated.

Filtering is a completely different process. After honey is extracted, it is rapidly heated, usually in flash heaters. These are tanks or pans filled with hot water or oil surrounding the pipes containing the honey that is moving through the tank. The honey is heated to the temperature of the water or oil, then pumped out of the tank and through very fine filters under pressure. Warmed honey runs through the filters much faster than even lukewarm honey does, so the process is fast and efficient. Once the honey is filtered, more sophisticated systems pump the honey through a similar system; however, rather than warming the honey, it is rapidly cooled. These are called flash coolers, and they do help maintain some of the volatiles, enzymes, and aromas that untouched honey has. Still, it's not the same when the process is complete.

Crystallized Honey

With incredibly few exceptions, all honeys crystallize sooner or later. Some much sooner. Some much later. How you handle crystallized honey says much about your concern for maintaining the quality of the product you and your bees have worked hard to produce. In dealing with this situation, beekeepers have, over the years, devised a multitude of ways in which we can do incredible harm to this delicate product. All of them involve heating the honey too hot, too fast. But you will have to deal with this, so invest some time, effort, and energy in figuring out a way to uncrystallize honey without damaging it. Here are some thoughts.

First off, there are two factors that hasten crystallization in stored honey: temperature and any impurities in the honey. The optimum temperature for honey crystallization is right about 57°F (14°C). Warmer or cooler, and the process proceeds more slowly. So storage location can be important. A room-temperature storage facility will slow this process, as will storage at freezing temperatures. If either of these is possible, certainly aim for the colder temperature. The not-too-serious drawback of honey becoming somewhat darker in colour when stored at room temperature should be considered if it is to be there for a long period of time, say six months or more.

The Drawbacks of Straining

Even a finely strained honey will probably have tiny particles remaining in it. This is precisely what gives it its unique flavor, colour, and aroma. But those tiny particles serve as the starters for the glucose crystals to grow on, and the whole process builds from there. Removing as many of these particles as possible reduces the likelihood of crystallization beginning. This is precisely why commercial honey packers filter their products to death; they do not want any particulate matter in a jar that may sit on a grocery store shelf or in a grocery store warehouse for months. Crystallized honey doesn't sell and they lose money. So it's a toss-up on how much of that material you want to remove. If you remove all of it, flavour, aroma, and quality are reduced. If you remove none of it, you maintain all the good qualities of the product, but enhance the probability of crystallization. A happy medium must be reached in your operation to satisfy both your customers and yourself.

Honey that has been extracted, strained, and poured into a storage container will begin, rapidly or slowly, to crystallize. Once crystallized, no further harm can come to it. The honey will sit like this for months. If it is at room temperature, it will very slowly darken; if it is stored at cooler temperatures, this darkening will be slowed considerably. It is only when you decide you no longer want a rock-solid pail of honey that things can start to go wrong.

Warming Crystallized Honey

To change crystallized honey to liquid honey, the temperature needs to be raised so the liquid in the matrix of glucose crystals and fructose solution (fructose sugar dissolved in water) warms enough to dissolve the glucose crystals, rendering the entire solution again a liquid. Essentially, sugar crystals are dissolving in warm water.

Filtering Exposé

Because the filtering system is closed—that is, the honey is under pressure, contained, and not exposed to the external environment—it is assumed that none of the aromas, volatiles, or other essential ingredients of the honey are lost. Anyone who has ever tasted honey before it is heated, pumped, filtered, and drained knows better. These systems change honey. Not much, but it is no longer the same.

Commercially filtered honey is crystal clear and devoid of any particulate matter. However, filtering diminishes the flavour, aroma, and quality of honey, which is why straining only is recommended.

Common instructions available on consumer labels say to "slowly" warm crystallized honey until it returns to its liquid phase. Do so in a pan of warm water, perhaps bringing the water to a boil, turning off the heat, and placing an open jar of honey in the water until it cools. Or keep the pan on very low heat, with the open jar in the pan. These techniques tend to not overheat the honey, but are impractical when considering a large plastic pail or an enormous drum. So what do you do?

Commercial honey packers have a technique that works well and isn't hard on the honey. The covers are removed from the large drums of crystallized honey and placed in a commercial-size oven, open side down over a grate. As the oven slowly warms the honey, it drips out of the drum, falls through the grate, and is captured below the grate in a tank that is part of the oven. Several drums can be placed in the oven at the same time so the blend of melted honeys is a consistently flavoured and coloured product. This is unfortunate, but the warming process is humane and the honey is usually not damaged because the heat is set at about 110°F (43°C) or so.

Barrel-Heating Band

Smaller-scale operators sometimes use a barrel-heating band that wraps around the center of the barrel and is heated with elements within the band. To liquefy an entire barrel, the band is set at a temperature considerably warmer than 110°F (43°C), and the heat dissipates throughout the rest of the honey in the barrel. In the end, the average temperature is 110°F (43°C), but inside the barrel close to the band the honey is considerably hotter, damaging the honey. Away from the band the honey is considerably cooler, probably not damaging the honey, but not becoming fully liquid either. Bands, for barrels or plastic pails, are not recommended because of the potential damage they can cause. Heating elements that you can put in a barrel are available, but these also tend toward too hot near the element, potentially damaging the honey.

Warming Boxes

Warming boxes work for both barrels and pails, and for most medium to small operations, these are common. Five-gallon (19 L) pails are common storage containers, and a warming box for these is easy to construct. Begin by making a wooden platform large enough to hold two to four pails so that they sit at least 1 foot (30.5 cm) apart,

Creamed Honey

HIVE HINTS

Creamed honey as a consumer product is much more popular in Europe than in the United States. However, it is slowly gaining in popularity in the U.S., especially as products with dried fruits and flavourings mixed in.

Creamed honey is simply crystallized honey, but the process is controlled by the beekeeper rather than left on its own in a pail somewhere. The process was perfected by E. J. Dyce from Cornell University in 1935, and is called the Dyce method.

Well-made creamed honey is smooth, creamy, and easily spread, like soft butter. It can harden, and if it does, it should be softened at a warm room temperature. It can be too soft, in which case it should be remade before being sold because it will not recrystallize with the same texture and crystal size.

Dyce Method

To make creamed honey using the Dyce method, the honey must be pasteurized to remove the naturally occurring yeast cells. These yeast cells will grow in an environment with increased moisture, and this process eliminates any chance of fermentation after the honey is crystallized.

To pasteurize, heat the honey to 150°F (66°C). This incurs some damage to the honey, but not very much, but fermentation would ruin it. As it heats, the honey should be stirred constantly to avoid hot spots in the pan, but not so vigorously as to incorporate air bubbles. When heated, strain it thoroughly through fine cloth to remove wax and other particles. Then, as rapidly as possible, reduce the temperature to about 70°F (21°C).

That's when starter is added. Starter is an already-creamed honey, one that is smooth, creamy, and with no noticeable crystals when you put some in your mouth. A good variety can be procured from a store or other beekeeper and use that the first time. You can use your own after that.

Mix in the starter honey with your now-cooled honey at the ratio of 1:10, starter:honey. Honey that is at 70°F (21°C) will be thick, and a good way to stir it is to get one of the creamed honey beaters from a beekeeper's supply company that fastens to your electric drill. Mix thoroughly until all of the starter is incorporated into the honey, but again, avoid incorporating air into the mixture.

When mixed, let settle for a few hours at room temperature to let any stray air bubbles or other foreign material rise to the surface, then skim if needed. Finally, pour into containers, seal, and allow to set for a couple of weeks at 57°F (14°C). One trick to consider is to fill the containers, then turn them upside down for the duration. This allows any lingering air bubbles or other material to "rise" to the surface, which will be the bottom. When complete, turn the jar back upright, and the top will be spotlessly clean and smooth. The final product should be butterlike, easy to spread, and smooth as glass. Once you have mastered creamed honey, try adding dried fruit, cinnamon, chocolate, or essential oils.

When Good Honey Goes Solid

Almost all honey will eventually crystallize. Honey in glass jars can be reliquified in a warming box. Plastic containers are more challenging, since the temperature needed to melt the honey will distort the plastic container.

The economics of using glass or plastic containers comes into play. Glass's weight will add to shipping charges and to the amount of work simply moving cases. However, lighter-weight plastic, though less expensive to purchase and to ship, makes warming crystallized honey problematic. Glass can be safely warmed; plastic cannot.

If you are selling your honey at the retail level, you should include with your packaging a small instruction statement on reliquifiying honey, but be sure to include information about heat and plastic. However, more and more beekeepers are ignoring those instructions because of the structural and chemical problems encountered when warming plastic containers.

though 18 inches (45 cm) is better, and there is at least 2 inches (5 cm) clearance between the pail and the outside edge of the platform and 4 to 6 inches (10 to 15 cm) in clearance above the lids. Raise the platform off the floor 4 or 5 inches (10 or 12.5 cm) to allow the wiring for one or more light sockets. Locate the light sockets on the platform to equally distribute the heat they generate. Make sure the pails are not too close to a bulb, so overheating doesn't occur (8 to 10 inches [20 to 25.5 cm] is good with a 40-watt bulb, further for larger-watt bulbs).

From here, you can get very creative. At its simplest, set your pails on the platform and measure the volume needed to make a box from 2-inch-thick (5 cm) Styrofoam insulation, leaving the appropriate side and head space, with the bottom resting on the floor, or, if the platform is large enough, on the edge of the platform floor. Glue the box pieces together and you're done. Plug in the 60-watt bulb, place the two pails in the box, put the top on, and leave it alone for six to twelve hours. Come back, turn the pails 180 degrees so the other side is exposed to the heat of the bulb, and by morning, the pails will be liquid.

The creative part comes in by adding a timer, a thermostat that turns the bulbs on and off when the interior reaches a preset temperature, small fans to circulate the interior air, and so on. These and other add-ons can be fashioned to automate the process and protect your honey at the same time. The key is to not let the honey temperature rise above 100°F (38°C), and to keep it that warm only as long as absolutely necessary to get it liquid.

If your only goal is to get the honey soft enough to remove it from the pail, warming won't take very long. The honey only needs to reach the slurry stage to pour easily. From there, it generally goes into a warmed bottling tank where it can continue melting, but in a very controlled environment.

If the honey is to be bottled directly from the pail, it will likely need to remain in the hot box longer. Be careful the honey doesn't overheat. Stir on occasion, use fans and thermostats, and turn the pail partway through the process.

A simple warming box consists of a platform, lightbulb, and Styrofoam box. Five-gallon (19 L) pails are placed on the platform, the light is turned on, and the box covers it all. With a single 40-watt bulb, crystallized honey is warmed soft enough to pour into a bottling tank in about ten to twelve hours, without overheating in the process.

☞

The Styrofoam box is covered with industrial strength aluminum foil (for durability). When the honey pails are placed on the platform, above, the light is turned on and the box placed over the honey pails, resting on the platform. The Styrofoam box doesn't need to be extremely durable or of heavy duty construction (and definitely not heavy because you will be lifting it a lot). A well-insulated box is what is most important.

☞

Two examples of water-jacketing bottling tanks are shown here. Water-jacketed bottling tanks can be precisely controlled to finish warming the honey. Do not leave honey in heated tanks after bottling, and remember to drain the water if the tank is kept in an unheated building when not in use.

To liquefy honey that has crystallized in retail containers—bears, bottles, and jars—beekeepers have been very creative. Because immersing a jar with a label in warm water is impractical, beekeepers will often design some type of hot box arrangement. An old refrigerator is common. A heat source (usually a lightbulb) is added, a small fan is used to circulate the air, and a thermostat is installed to control the temperature, and presto! You have a hot box for jars. The construction cost is minimal, the electric draw is minimal, the footprint is tiny, and the return is significant.

Warming plastic containers, however, can be problematic. Subject them to a cooler temperature, and the honey will take a very long time to melt. But keep it above a certain temperature, and the honey will melt but the container will become distorted and give the honey an off-flavor. Dealing safely with crystallized honey in retail plastic containers is difficult.

Bottling Tanks

There are a variety of good bottling tanks available for bottling honey. They all use the same design concept, which is that honey is held in a tank that sits inside another tank. There is a water jacket that surrounds the honey tank inside. The water can be thermostatically set to hold at a temperature you choose. Most have an external glass pipe that shows the amount of water in the heating tank, so it doesn't go dry and you know when to refill it. The concept is ideal, but in reality, the system is far too often abused. How? It's easy: honey that has crystallized is softened in a heating box similar to that described above. It is carefully warmed until soft enough to pour, and then it is poured into the tank for bottling. There, it is warmed to a handy bottling temperature, say 80° to 90°F (27° to 32°C) or so, and gets bottled. If all of it isn't bottled, the remaining honey stays in the tank, "just in case." There it stays for a day, two days, a week, or more. Honey that is constantly kept at that temperature will slowly darken, slowly degrade. The lesson here is to soften the honey in the warming box, pour it into the bottling tank, warm further to the bottling temperature, and use it all, even if you have to put what's left back in a pail. Don't leave it in the tank.

A plastic pail with a spigot: the simplest, easiest, and least expensive bottling operation. You can use the pail for other purposes, such as mixing sugar syrup, it stores easily, and it can be replaced easily.

Fueled by Frustration

Honey production (and its marketing and selling) can be a humbling experience for a seasoned beekeeper. Deep down, this is the best test of your passion for beekeeping and the future of your business. In fact, this frustration will only drive you on, and in the long run it will make you appreciate even more the value of the product you are now going to sell.

Honey harvested and bottled for home use can be stored in any number of resealable containers. Honey for sale, however, requires additional considerations.

FOR SALE: HANDMADE HONEY

New beekeepers certainly can produce high-quality honey, but more likely they are still focused on mastering the basics—learning to keep their bees healthy, the seasonal routines of the hive, and so on. If you've mastered those basics, you've been around a while and have acquired the skills necessary to maintain strong colonies, to routinely produce queens and make divides, to manage bees for optimum honey production, and especially to make the income necessary to grow past the hobby stage. (Besides, if you've come this far in this book you should now be past the point of producing a simple commodity—honey.) From here onward, what you produce and sell is that rare and special product of fine crafted honey.

Producing quality, handmade honey requires more effort than producing the basic end-of-the-season, put-it-in-a-pail product. Once you consider the new skills you've had to master, the additional equipment you've had to purchase, and the extra time required to do the job right, you might feel like you are back to square one again.

Consistently producing varietal honeys makes you a better beekeeper. You become more aware of your bees and the floral environment all the time. You take better care of your bees not just during the early spring months but all season long. As a result, you're harvesting more often and going through all the pre- and post-harvest chores required to do this exactly right each time. In short, you are working harder to produce the crops that you intend to sell, but you are producing better crops.

Great honey producers are not necessarily great salespeople, becoming which requires another set of skills to cultivate. And now you are selling a new product from your standard honey: it's the difference between a mid-winter tomato in the grocery store and one that's still sun-warm ripe and just picked from the garden.

You will likely be using different containers and different labels, and the honey will be priced to reflect the additional cost incurred while producing the new product. New sales outlets and customers will have different shelf requirements. Let's examine marketing techniques that address these issues before you begin.

Upgrading the Bottles

The fine crafted honey product should have a container that is different from the generic product. The label will help differentiate the two product lines, but the container should also reflect the handmade honey's unique quality. There are dozens of container styles available that are made (and sold) in quantities that keep their prices reasonable, but this is not the time or place to scrimp: a wonderful product in a cheap container looks like a cheap product, period.

Consider choosing a new shape of glass jar. If using plastic, consider switching to glass or upgrading to a more useful plastic shape. Observe what else is available on the honey shelf in stores and at farm stands. If every bottle is in a glass cylinder, the decision of how best to differentiate the quality product becomes easier. Aspire to a memorable look that will help customers find your jar next time, and every time.

Designing Appealing Labels

Labels chosen and designed for honey jars are essential for identifying not only the honey's quality but also its origins. Observe other honey labels in the local market, study labels from companies that advertise in the beekeeping journals, and then examine the labels of other specialty products (such as wine, sauces, oils, and other gourmet items). All promote an atmosphere that immediately alludes to the quality of the products.

The artisan honey label must have certain information on it to meet legal requirements: name, address, contents, weight (never handwrite the weight on a label), and other information, including a bar code and a nutrition label. Leave some room for romance, poetry, and singing the praises of this particular honey. If you use a single label for all of your varietal honeys, a label at the back or on top should identify each individual honey variety.

Playing The Name Game

Naming your varietal honey is easy—certainly much easier than producing it—Blackberry, Locust, Starthistle, and so on. But a quality, handmade honey requires a bit of thought, and possibly copywriting. The names could be seasonally influenced: Spring Blossom, Solstice Flower, Autumn Harvest. A one-time-harvest crop could be called One Thousand Flowers.

Inverted containers that keep the honey against the jar opening are convenient and offer ample room for affixing prouct labels.

These traditional honey jars are known as Classic Gambers. If all other honeys on the shelves are bottled in these containers, consider differentiating your product with another style.

When considering new containers, evaluate their material, shape, size, and where the label will be placed. The lid should also be considered for both its appearance and whether it can accommodate a label.

Don't forget to boldly identify where the honey comes from. Local honey is always better than honey from an unknown source (especially if it's imported), so make sure everybody knows it is local. List the name of the county, province, or even city from which it was harvested. (Clever beekeepers have even given honeys from different beeyards different number names to correspond to their telephone area codes or postal codes.)

Finding Retail Outlets

Special honeys should be sold in special places. Regular places, too, but look for special places. High-end gourmet food stores are great. Places where discerning chefs and cooks shop are perfect. Think gourmet food sections that are in most grocery stores now. What about a special booth at a farmers' market (different from your regular booth), where you sell only your special honey. These are locations that aren't usually associated with "regular food". Save the hardware stores, farm stores, and general grocery shelf for your regular honey.

However, at the places you have been selling honey consider adding these as a new line. It will require additional shelf space and that always poses a problem. A compromise that often works is to divide your original space and combine the two products ... but separate them on the shelf as much as the store owner will allow. If both sell well, you can bet the shelf space will be found to accommodate a new variety. If they don't, then you know you have to find other outlets.

Pricing Honey for Profit

Many beekeepers find the pricing aspect of this the most difficult part of sales. How much do you charge? How much for wholesale, retail, bulk, and so on? There are quantity discounts, regular discounts, and no discounts—all factors need to be considered. All pricing decisions boil down to two things: how much does it cost to produce and how much profit do you need to stay in business?

The cost analysis is the most difficult to create because agriculture is such a system of variables: no two years are alike, no two crops are equal, and input costs vary.

Input costs are easy to calculate with good record keeping: fuel, medications (optional), new equipment to amortize, bottles, labels, vehicle expenses, replacement bee costs, and more. But quantifying your own labour is a challenge, and too often beekeepers think they can't afford to pay themselves a wage to add to the cost equation. Here is an example that helps calculate the cost to your business for using your skills, labour, and time.

Imagine that you are returning home after the first inspection in the spring. Your foot slips off the back of the truck while unloading supers and you land hard on your left ankle, which shatters. After surgery, you are completely immobilized for six weeks and on crutches for another four months. Your personal contribution to the honey harvest is going to be minimal and you will need to hire help to stay in business, to keep your bees alive, and to continue to supply your customers with the honeys they have come to expect.

If you're lucky you have family to fill in, but imagine that you don't. How much would you need to pay one, maybe two people to do as good a job as you do, all season long? That figure, at minimum, is the cost of your labour to your business. And that is the wage you should be paying yourself for doing all this work every season.

When figured into all other costs you may still decide not to pay yourself and to keep the money in the business. That's a business decision; a real-world price for your honey should account for the real-world costs of the decision.

Many small business owners consider that unpaid wage their profit, which is one way to calculate that figure. But if you have to put food on the table* with the earnings from your honey business, you will be more considerate of how much you pay the help, and you will likely pay them better.

The Bottom Line

Producing, then selling, varietal honey is a challenge because it requires special skills to produce, and it is priced higher than other honeys in the marketplace. If you are even slightly successful you must add the cost of acquiring and harnessing those skills to your list of costs and charge accordingly. Remember: these honeys are better than the commodity honey on the grocery store shelf, and you deserve the respect and profits they bring.

*See chapter 5, page 120, for some ideas.

The lid on your container can make it easy to identify the honey inside simply by glancing at the color or style of the top outside.

Early season management consists of attending to the details. Knowing the biology of the beehive and anticipating its needs, making sure the health of your bees is excellent, knowing the surrounding floral resources, and being prepared to accommodate the influx of nectar and honey are key. Also be aware of the weather and growing degree days, and making the time to do the right things at the right time in the right way—every time.

CONCLUSION: THE WISE BEEKEEPER

Not everybody can keep bees. Fewer keep bees well. And even fewer not only manage their bees in good fashion but also consistently produce varietal or artisan honeys. These are, indeed, rare beekeepers.

I trust by now you know why. Managing honey bees well requires attention to the details of honey bee biology and doing the right thing, the right way, at the right time, every time. And all the while knowing that what is best for your bees is not always best for you, but nevertheless always doing what's best for the bees.

Producing a honey crop also demands putting the bees first in your production scheme. In addition, you must keep tabs on the weather, and you must know of the thousands and thousands of different plants that grow where your bees live. Not only that, but you must also know the complicated and demanding life cycles of these plants—when do they bloom and what else is blooming at the same time? Plus, you must be aware of their temperamental requirements to produce their honey bee bribes in season. And when they do, are your bees visiting those plants, and is that really the honey you think it is? You can't guess, ever.

If your colony remains relatively healthy and has a respectable population when the various honey flows begin, and if the weather cooperates, you will, most years, be able to produce a decent honey crop while leaving more honey for the colony than they will need.

But it's a special beekeeper who does all of this and more to produce that unique and rare crop. This is the mark of not only a good beekeeper but also a culinary artist. And a picky artist at that. Not all honeys deserve this much attention. Some, and you need to know them, are acceptable but not fit for finer dining, while others are barely pedestrian in flavour and texture. But those chosen few, oh, they are the best you can imagine.

Varietal? Handmade? Only the uninformed do not truly appreciate these honeys. A true handmade honey is rarer and more special than any varietal honey because only handmade honeys are a once-in-a-lifetime blend—no spring bouquet is like any other springtime mix, from here or from any other place. No summer wildflower field is ever the same again, and the blend of autumn's best is only this year's vintage, for next year will be different.

Of course, it goes much further than simply having enough healthy bees in the right place at the right time to produce these wonderful crops. The bees must be coaxed from their hard-earned bounty with respect, with care, and without damage. And the wise beekeeper knows too well that to be greedy in the midst of plenty will mean trial and tribulation for the bees down the road. There is a broad and easily not-crossed line here, for it should never be feast and famine, but always feast and feast. All good beekeepers leave a generous and healthy portion for the bees.

Next, the crop must be carefully delivered unsullied to the place of preparation. Once arrived, it should be ever so slightly warmed, if it isn't already warm from the hive, to make the next process efficient and reduce its exposure time to the elements of the world. This is honey at its most vulnerable, and it must be protected.

Easing honey from the comb into the containers we want to keep it in can be a dangerous process, too, especially when heat enters the equation. The good beekeeper uses just enough warmth to hasten the process but not in the slightest harm the honey during this journey.

Uncappers of all sorts, radial extractors, sumps and pumps, tanks and heaters, coolers and pipes, spinners, bottlers—all of these portray an image of factory-farm production. And sometimes, unfortunately, that is the case.

But handmade honey by its very nature does not project that tarnished image. Rather, it is a homemade image, one of skilled and patient hands offering tender care for the animals that magically produce this marvelous sweet. It is the image of calm and practiced shepherds guiding this delicate crop from bee to bottle.

If you follow the guidelines here, you will always be working with your bees toward that common goal—to produce handmade honeys truly worthy of the name. And when you are successful, you can honestly say that you stand tall among the Brotherhood of Better Beekeepers.

And you and your bees will thrive.

Management later in the season is more 'honey' management than 'honey bee' management. Making the right choices—not harvesting uncapped honey or frames with pollen or brood is an essential right choice. Checking honey moisture before harvesting is a critical step for efficient use of your time and only harvesting the perfectly aged honey. And finally, protecting your harvested crop and keeping it safe all the way from bee to bottle is paramount when producing a fine handmade, or varietal honey.

Chapter 5

Using What You and the Bees Have Made:
Varietal and Handmade Honey Recipes

MAKING INFUSED HONEY

Infusing Honey with Fresh Herbs

You can add a hint of fresh herb flavour to your light and medium honeys. First, gather the herbs as early in the morning as possible while they're still fresh and full of flavour. Wash thoroughly under cold water to remove soil, spent flowers, and old leaves. Almost any herb can be used. Consider delicate-flavoured herbs such as rose petals, chamomile, or lavender, or stronger herbs, such as rosemary, anise, or mint.

Use a light, mildly flavoured spring honey. Once you've filtered your crop, fill pint jars about four-fifths full so adding the herbs won't create overflow. Purchase reusable tea bags or make your own from fine mesh nylon from a fabric store. The bags should be 2" to 4" (5 to 10 cm) square with either a drawstring or a clamp on the opening.

Chop delicate-flavoured herbs into moderate to fine pieces and add 3 to 5 tablespoons to the mesh bag and seal. For stronger herbs, use 2 to 4 tablespoons.

Place a bag in each jar and set in a sunny windowsill for 1 to 2 weeks. After the first week, taste the honey, and if strong enough, remove the bag. If more flavour is desired, either add more herbs or let the herbs steep in the sun for another week. After the flavour has reached its peak, remove the honey from the windowsill, carefully remove the bag of herbs, and discard it. The honey can then be rebottled in gift jars or to sell.

Warmed Infused Honey

You can, of course, speed up this process, and at the same time have better control of the intensity of the flavour you add to the honey. This procedure requires a larger quantity of herbs, and heating the honey to 180°F (82°C). This will, to a degree,

CRUSHING POTENT HERBS

For rosemary, for instance, crush the rosemary leaves to release the flavor. Place in a plastic zip-top bag and roll over them with a rolling pin one way, then turn the bag 90 degrees and roll the other way, being careful to crush them just enough to release the flavor. If you hold the open bag to your nose, there should be a strong, but not overwhelming, fragrance of the herb in the bag.

allow some of the aromatic properties of honey to be released, and some enzyme levels will be reduced, but neither in significant amounts. You are adding additional fragrance with the herb, so that will help compensate.

To make controlled infused honey, use a double boiler. Add 2 cups (680 g) of honey, using a light, mild variety, so the infusion flavours are not masked. To the honey, add the herb of your choice directly, or use a mesh bag to contain them and reduce fragments in the finished product (as described above).

Add the crushed herb to the honey as is or in a mesh bag and bring the mixture to 180°F (82°C). Continue to heat at this temperature for 10 to 12 minutes. That should be long enough to release as much of the flavour that amount of herb will have available, while not harming the honey. When done, strain (if needed), place in a marked jar, and cap.

You can use almost any herb you can imagine, including any of the many mints (use 2 cups [192 g] of crushed mint leaves), chives or garlic chives (use 1 cup [48 g] chopped), garlic (use 2 cloves minced), citrus zest (use the zest of a whole orange, tangerine, or grapefruit), oregano (use 1 1/2 cups [96 g] crushed leaves), thyme (use 1 1/2 cups [58 g] crushed leaves), or parsley (use 2 cups [120 g] crushed leaves).

Infusing Honey with Flowers

You can also infuse honey with flowers, but what you are infusing will be primarily the nectar flavour, which is very delicate. Suggested flowers are chive blossoms, violets, clover blossoms (use different varieties for slightly different flavors), black locust blooms, honeysuckle blossoms, autumn olive or Russian olive blooms ... the list goes on and on. Try any flower that has a pleasant, aromatic fragrance.

You will need to experiment with blossoms more than with leaves because the flavours are so delicate, the amount of nectar is often minute, and the heat will drive off some of the aromatics. Don't crush the blossoms before adding them to the honey. Rather, pick them as early in the morning as possible on the day that they open, before bees, the weather, or time reduces the amount of nectar and fragrance. Remove as much of the stem as possible, and any green sepals, leaving just the petals and the reproductive parts of the blossom.

Follow the same procedure with the mesh bag, using about the same volume of flowers as you did of the leaves. Bring the heat up to 180°F (82°C), but then immediately remove from the heat. If you need to increase the intensity, instead of cooking longer, remove the original flowers and add new flowers when the honey has cooled, and reheat as before.

The Honey Palette

Black Locust

Basswood

Orange Blossom

Wild Raspberry

Clover

Alfalfa

Wildflower

Blueberry

Dips and Dressings

The following dips and dressings were inspired by recipes from The (U.S.) National Honey Board. Varietal and handmade honeys were incorporated, and herbs and spices for flavour, texture, or colour were added. Enjoy! (And remember, serving sizes are all relative. How hungry *are* you?)

Island Dip

- ¹/₂ cup (120 g) ketchup
- ¹/₄ cup (85 g) honey (a sweet, light, early spring variety, such as black locust, maple, or even clover)
- 2 tablespoons (30 ml) lemon juice
- 1 teaspoon cornstarch
- ¹/₂ teaspoon garlic salt
- ¹/₂ teaspoon onion powder

In a microwavable bowl, combine all the ingredients and mix well. Microwave on high for 2 to 3 minutes, pausing every 30 seconds to stir. Cook until the mixture boils and thickens. Cool. Serve with chicken or vegetables.

Makes about ³/₄ cup (235 g)

Sweet Cheese Dip

- 8 ounces (230 g) cream cheese
- ¹/₄ cup (85 g) honey (a light, mild variety, such as clover)
- 1 ¹/₂ (7.5 ml) teaspoons vanilla
- ¹/₄ teaspoon cinnamon
- 1 teaspoon (5 ml) lemon juice
- 1 teaspoon (5 ml) lime juice
- 1 teaspoon (5 ml) orange juice

In a medium bowl, beat the cream cheese until light and fluffy. Add the honey, vanilla, cinnamon, and juices and mix very well. Serve with fruit.

Makes about 1 cup (335 g)

Sweet 'n Smooth Curry Dip

- 1 cup (240 g) mayonnaise
- ¹/₄ cup (85 g) honey (a medium strong variety, such as a fall sweet goldenrod or actinomeris)
- 1 tablespoon (6 g) curry powder
- 1 tablespoon (15 ml) white wine vinegar
- ¹/₂ teaspoon (1.5 g) garlic powder

In a small bowl, combine all the ingredients and mix well. Refrigerate for at least an hour to allow the flavours to blend. Serve with raw vegetables.

Makes 1 ¹/₄ cups (340 g)

Hot Honey Dip

- ¹/₂ cup (170 g) honey (a medium-flavoured variety, such as a midsummer wildflower)
- ¹/₂ cup (120 g) Dijon-style mustard
- 1 teaspoon (5 ml) Worcestershire sauce
- Freshly ground black pepper, to taste
- Cayenne pepper, to taste

In a small bowl, combine all the ingredients and mix thoroughly. Use as a dip for flavoured bread or fresh or fried vegetables (or even deep-fried cheese). This is a hands-down favorite for chicken wings.

Makes 1 cup (290 g)

Very Pink Dip

Because the sweetness of the strawberries will vary, you may need to adjust the amount of honey you use in this recipe. Start with 3 tablespoons (60 g), and then add more, if needed, until you get the taste that's just right.

- 1 cup (230 g) plain yogurt
- 3 tablespoons (60 g) mild-flavoured honey
- 1¹/₂ cups (300 g) mashed strawberries, with juice

In a medium bowl, blend together all the ingredients, stirring until well mixed. Cover and chill until needed. Serve with fruit.

Makes 2 cups (590 g), or 1 great breakfast

Honey Mustard Dip

Your choice of mustard will lend a distinctly different flavor to the dip.

- ¹/₂ cup (170 g) honey (a medium flavourful variety, such as alfalfa or a summer wildflower)
- ¹/₂ cup (120 g) prepared spicy brown mustard (or Dijon mustard)

In a small bowl, combine the honey and mustard. Serve with chicken or vegetables.

Makes 1 cup (290 g)

Honey Mustard Dressing to Die For

- 1¹/₄ cups (300 g) mayonnaise
- ¹/₃ cup (115 g) summer wildflower honey
- 1 tablespoon (15 ml) vinegar
- ²/₃ cup (160 ml) extra virgin olive oil
- 1 teaspoon (3 g) finely chopped onion
- 2 tablespoons (8 g) finely chopped fresh parsley
- 2 tablespoons (30 g) prepared mustard

In a medium bowl, add all the ingredients and mix together with a large whisk until smooth and creamy. Use for salad greens, or with chicken or seafood salads.

Makes about 2 cups (450 g)

Tom's Teriyaki Dipping Sauce

Chef Tom Brotherton shared this recipe just for this chapter.

- 10 ounces (280 g) honey (choose a light, mild spring blend)
- 15 ounces (440 ml) teriyaki sauce
- ¹/₂ teaspoon ground ginger, or 1 tablespoon grated fresh ginger
- 1 cup (235 ml) orange juice, at room temperature
- 1¹/₂ cup (355 ml) water
- ¹/₂ cup (65 g) cornstarch

In a medium mixing bowl, blend together the honey, teriyaki sauce, ginger, orange juice, and 1 cup (235 ml) of the water, mixing thoroughly until the honey is well dissolved. If using ground ginger, make sure that it does not clump. Pour into a medium saucepan over medium heat and slowly bring to a boil. Do not hurry this step. Meanwhile, in a cup, mix together the cornstarch and remaining ¹/₂ cup (120 ml) of water. As the sauce warms, thicken it to taste with the cornstarch mixture. When the sauce boils, reduce the heat and simmer below a boil for 5 minutes. The desired thickness for dipping is when it coats the back of a spoon. Almost anything dipped in this special sauce tastes better.

Makes 3 cups (1,375 g)

Herb Dip 'n Dressing

- 2 cups (460 g) sour cream
- 6 tablespoons (120 g) honey (a light, mild variety, just for the sweetness)
- 2 tablespoons (36 g) frozen orange juice concentrate, undiluted
- 2 tablespoons (30 g) Dijon mustard
- 2 teaspoons (10 g) cream-style horseradish
- 2 teaspoons rosemary, crushed
- 1 teaspoon chervil, crushed
- 1 teaspoon basil (pick the variety you like best), crushed
- 1 teaspoon oregano, crushed
- $^1/_4$ teaspoon salt
- $^1/_2$ teaspoon onion salt
- $^1/_2$ teaspoon white pepper
- $^1/_4$ teaspoon garlic powder

In a medium bowl, combine all the ingredients and mix thoroughly. Cover and refrigerate for several hours to blend the flavours. Stir before using. Use as a dip for chips, shrimp, or vegetables, or as a dressing for a green salad.

Makes 3 cups (675 g)

Easy BBQ Sauce

For best results, don't use a sweet onion—find one with some punch and character. Also, the darker the varietal honey used, the heartier this sauce will be.

- 1 medium onion, finely chopped
- 1 tablespoon ground coriander
- 2 tablespoons (40 g) honey (dark varietal, such as orange blossom or tulip poplar)
- $^1/_4$ teaspoon freshly ground black pepper
- $^1/_4$ –$^1/_2$ teaspoon garlic powder
- $^1/_8$ teaspoon cayenne pepper
- 3 tablespoons (45 ml) lemon juice
- $^1/_4$ cup (60 ml) soy sauce

In a saucepan, combine all of the ingredients over moderate heat and simmer for 5 to 10 minutes, until thickened as desired. Do not bring to a boil. Do, however, taste now and then and add additional spices (salt, cayenne pepper, garlic powder, or freshly ground black pepper) to taste before removing from the heat.

To use barbecue sauce on the grill: Cook meat about halfway through on both sides (by turning once before adding the sauce). Add the sauce evenly with a brush to one side. When the first side is finished cooking, turn the meat, add sauce to the second side, and finish cooking. This is a stand-up-and-be-noticed barbecue sauce because of the pepper and onion, but a strong honey won't get lost in the mix and adds a smoothing (not soothing) effect to the sharp cayenne pepper flavor.

Makes $^1/_2$ cup (125 g)

Honey Board Barbecue Sauce

This recipe was one of the first put out by the National (U.S.) Honey Board in the late 1980s. It's a standard today.

- 1 can (15 ounces [440 ml]) condensed tomato soup
- ¹/₂ cup (170 g) honey (use a strong, flavourful late summer wildflower variety)
- 2 tablespoons (30 ml) Worcestershire sauce
- 2–3 tablespoons (30 to 45 ml) salad oil
- 1 tablespoon (15 ml) freshly squeezed lemon juice
- 1 teaspoon (5 g) mustard (yellow or Dijon)
- Dash cayenne or bottled hot pepper sauce (optional)

In a saucepan, combine all of the ingredients and bring to a boil. Immediately reduce heat and simmer, uncovered, for 5 minutes. Use as a barbecue sauce on the grill, being careful not to burn it (don't apply it too early in the process). It can also be used as a dipping sauce.

Makes 2 cups (500 g)

Sweet Curry Dipping Sauce

- 1 cup (230 g) sour cream
- 6 tablespoons (120 g) honey
  (a dark, strong varietal, such as buckwheat)
- 2 tablespoons (30 ml) cider vinegar
- 2 teaspoons (13 g) curry powder
- ¹/₄ teaspoon ground cumin
- ¹/₄ teaspoon garlic powder
- ¹/₄ teaspoon salt
- 2 to 3 tablespoons (8 to 12 g) chopped fresh parsley

In a small bowl, combine all the ingredients except the parsley and blend well. When blended, stir in the parsley. Cover and refrigerate. Serve with vegetables or flavoured crackers.

Makes 1 generous cup (400 g)

Salads

Sweet Summer Salad

- 1 cup (155 g) cubed watermelon
- 1 cup (155 g) cubed cantaloupe
- 1 cup (155 g) cubed honeydew
- 1 cup (145 g) blueberries
- Juice of l lemon
- $^1/_2$ cup (170 g) honey (a light and mild variety)
- $^3/_4$ cup (45 g) fresh whipped cream

In a large bowl, mix the watermelon, cantaloupe, honeydew, and blueberries. Dribble the lemon juice all over the fruit, then drizzle with the honey. Stir just enough to get both the lemon and the honey on most of the fruit. When mixed, refrigerate for at least an hour. Just before serving, fold in the whipped cream. Serve in a clear glass bowl.

Variation on the whipped cream: Chill the cream before whipping. Add $^1/_2$ teaspoon (2.5 ml) vanilla and 1 tablespoon (20 g) of the same honey used in the salad. Whip until you get soft peaks, and then keep chilled until needed.

Makes 8 servings for normal people, but not enough for 4 teenage boys

Applesauce Garnish

This is delicious warmed up and used as a garnish for white meats. Serve it cold as a garnish for vegetables.

- 3 apples (use a tart, not too juicy variety, such as Granny Smith)
- $^1/_4$ cup (60 ml) orange juice (with pulp)
- $^1/_4$ cup (85 g) honey (a mild, early spring variety, such as locust or red bud)
- $^1/_2$ tablespoon (7.5 ml) lemon juice (to preserve apple color)

In a blender, combine all the ingredients. Blend to desired consistency, without pureeing completely.

Makes enough for 1 large pork roast, or a side garnish for lunch

Waldorf Salad with Chicken

- 2 tablespoons (40 g) honey (a mild, spring variety, such as any spring tree or berry honey)
- 1 tablespoon (15 g) Dijon mustard
- 1 tablespoon poppy seeds
- ¹/₂ teaspoon grated lemon peel
- ¹/₃ cup (80 ml) freshly squeezed lemon juice
- ¹/₃ cup (80 ml) salad oil (or extra virgin olive oil)
- 3 cups (330 g) bite-size pieces cooked chicken or turkey (moist breast meat or tender thigh meat)
- 1 apple, unpeeled but cored and diced
- ¹/₄ cup (30 g) diced celery
- ¹/₂ cup (50 g) chopped scallions (use as much of the tops as possible)
- ¹/₂ cup (60 g) toasted sliced almonds
- Lettuce leaves, for serving
- Fresh parsley, for garnish

In a large bowl, mix together the honey, mustard, poppy seeds, lemon peel, lemon juice, and oil. Let stand at room temperature for a half an hour. Add the chicken and toss gently, then refrigerate until ready to serve. To serve, add the apple, celery, scallions, and almonds to the chicken mixture. Serve on lettuce-lined plates, and garnish with fresh parsley.

Makes 6 to 8 servings

Summer Mix with a Kick

Use high-quality fruit that is in season for this kicked-up summer fruit mix. Try peaches, apricots, strawberries, cherries, melon (watermelon, cantaloupe, honeydew, or honey rock), blueberries, raspberries, blackberries, or whatever else you can find, think of, or gather.

- 1 quart (0.95 L) summer fruit
- 1 teaspoon (5 ml) freshly squeezed lemon juice
- 3 to 5 tablespoons (45 to 75 ml) freshly squeezed orange juice
- 1 generous tablespoon (15 ml) Grand Marnier
- 1 ¹/₂ tablespoons (30 g) summer honey (such as black locust, clover, or basswood)

Cut peaches or apricots into thick slices; strawberries and cherries in half or quarters; and melons in ¹/₂-inch (1 cm) cubes or balled. Leave blueberries, raspberries, and blackberries whole. Prepare enough of any or all of these to fill your serving bowl about half full. Pour the lemon juice over the fruit. In a small bowl, combine the orange juice, Grand Marnier, and honey and mix well. Pour over the fruit. Gently toss, cover, and marinate in the refrigerator for a couple of hours to overnight. Serve with whipped or clotted cream.

Makes 4 servings

Vegetables

Layered Zucchini

To boost the flavor in this recipe, choose a gourmet bread made with cheese, onion, peppers, or garlic.

- ¹/₂ loaf artisan bread, cut into ³/₄ inch (about 2 cm) cubes (2 cups [100 g])
- 2 to 3 small zucchini, quartered and cut into 2 inch (5 cm) strips (about 2 heaping cups [230 g])
- Onion powder or garlic powder, to taste
- ¹/₂ cup (170 g) honey (choose a variety of medium flavour and color)

Preheat the oven to 350°F (180°C). Cover the bottom of a 9-inch (22.5-cm) baking dish with bread cubes, then add a layer of zucchini and sprinkle with onion or garlic powder. Drizzle a little honey over the top of the zucchini. Repeat layers until you are out of ingredients. (Depending on the thickness of the zucchini slices, you'll be able to make 3 to 5 layers.) Bake for about 30 minutes, or until the zucchini is tender, but not mushy.

Makes 4 or 5 servings

Honey Baked Beans

- 4 slices bacon, diced
- ¹/₂ cup (80 g) coarsely chopped onion
- 4¹/₂ cups (1,170 g) canned navy beans
- ¹/₂ cup (170 g) honey (a medium-dark, medium-flavourful variety, such as a summer or fall wildflower)
- ¹/₂ cup (120 g) ketchup
- 1 tablespoon (15 g) prepared mustard (or use more if you like mustard)
- 1 tablespoon (15 ml) Worcestershire sauce
- ¹/₂ teaspoon garlic powder

Preheat the oven to 350°F (180°C). In a small frying pan, cook the bacon over medium-high heat just until some of the fat is rendered. Remove the bacon to a plate lined with paper towels to drain. Lower the heat. Discard but save for later all but 1 tablespoon (15 ml) or so of the fat. Add the onion and cook until tender. In a 2-quart baking dish, combine the beans, bacon, onion, honey, ketchup, mustard, Worcestershire, and garlic powder, mixing to combine thoroughly. Cover and bake for 30 minutes, then uncover and bake for 45 minutes longer. Be careful not to overcook.

Makes 8 to 10 servings

Sweet Breaded Eggplant

- 1 large eggplant
- Salt
- Cracker crumbs (try a flavoured variety with a hint of cheese*)
- 2 eggs
- 2 tablespoons (30 ml) whole milk
- 1 cup (340 g) plus 2 tablespoons (40 g) honey (find the mildest and sweetest variety you can)
- 1 cup (225 g) butter, at room temperature

Slice the eggplant crossways in ¼-inch (6 mm) slices. Shake salt on both sides and put in a colander in the sink. Leave it there for 30 minutes, or long enough for water to drain. Thoroughly rinse the salt and any brownish juice off the slices. Dry with paper towels. Put the cracker crumbs in a flat baking dish. In another dish, barely beat the two eggs and add the milk and the 2 tablespoons (40 g) of honey.

Preheat the oven to 200°F (95°C). In a large frying pan, melt a few tablespoons of the butter over medium heat. Meanwhile, dip the eggplant slices in the egg mixture, then coat generously with the cracker crumbs, rolling the slice in the crumbs. Make sure both sides are well coated. Place the coated slices in the pan just when the butter begins to bubble, and fry both sides until brown and tender. Place the fried slices in an oven-safe pan and place in the oven to keep warm. Add additional butter, if needed, to the frying pan between cooking each slice or slices. While the slices are frying, add the remaining 1 cup (340 g) of honey and a tablespoon of butter to a coffee cup (or syrup pitcher) and place in the oven until the eggplant is ready to serve. Pour the honey sauce over the slices when serving.

*To make crumbs: Put crackers in a zip-top bag and use your rolling pin to crush. Make at least 1 cup (100 g) of crumbs, maybe 2 (200 g), depending on the size of your eggplant.

Makes 4 to 6 servings

QUICK TASTE

Sliced Tomatoes

Slice as many ripe tomatoes as needed, plus a not-too-large but ripe cucumber, and arrange them on a large platter. Sprinkle liberally with feta cheese, and dust with a sprinkle of Parmesan, Asiago, or Romano cheese. Tumble sliced ripe olives (sliced green olives stuffed with cheese, garlic cloves, or almonds make a remarkable alternative) all over the plate. Mix equal parts of a light, mild honey (such as black locust, sage, or clover), tarragon vinegar, and extra virgin olive oil. (The final mix should be about ½ cup [120 ml], adjusting more or less of any ingredient to taste.) Drizzle the honey blend over the plate. Finish with a handful of finely chopped fresh chives, or better, garlic chives. Garnish with fresh parsley.

Entrées

Sweet Chicken Quarters

- 3- to 4-pound (approx. 1.5- to 2.0-kg) chicken fryer, cut into quarters (or use various pieces)
- ¹/₂ cup (170 g) honey (choose any honey that is medium in colour and flavour, such as alfalfa, citrus, clover, or mild summer wildflower)
- 2 tablespoons (30 ml) soy sauce
- ¹/₂ cup (112 g) butter, at room temperature
- 1 teaspoon (3 g) cornstarch

Preheat the oven to 350°F (180°C). Place the chicken pieces in a shallow baking pan as flat as they will lie. In a small bowl, combine the honey and soy sauce. Brush the chicken pieces with the butter, and then drizzle on the honey mixture. Bake for 1¹/₂ hours, or until the chicken is tender, basting often with the sauce.

When done, remove the chicken to a serving platter. In the baking pan, make gravy by combining the drippings with any remaining honey mixture and the cornstarch. Serve over the chicken.

Makes 4 to 5 servings

Scallion Chicken

- 2 tablespoons (30 ml) extra virgin olive oil
- 3 pounds (1,365 g) chicken (a cut-up fryer, or any choice of parts)
- 2 tablespoons (30 ml) soy sauce
- ¹/₂ cup (120 ml) dry sherry
- 2 tablespoons (40 g) honey (use a fairly strong fall variety, such as goldenrod or even buckwheat)
- 1 teaspoon (6 g) salt
- 6 or 7 scallions, thinly sliced, using as much of the tops as possible

Heat the oil in a large skillet over medium-high heat. Add the chicken and cook, turning once, until browned, about 10 minutes. While cooking, combine the soy sauce, sherry, honey, salt, and scallions in a medium bowl. When the chicken is browned, drain the oil and add the sauce. Reduce the heat slightly and simmer, covered, for 30 minutes, or until tender. Pour the sauce over the chicken when serving.

Makes 2 to 4 servings

Honey-Glazed Duck

- 1 duckling (about 5 pounds [approx. 2.5 kg])
- ¹/₂ cup (170 g) honey (use a mild but flavourful variety)
- ¹/₂ teaspoon garlic salt
- ¹/₂ teaspoon onion salt
- 1 teaspoon (6 g) poultry seasoning
- ¹/₂ teaspoon paprika
- 1 teaspoon (6 g) salt
- ¹/₂ cup (120 ml) orange juice, with pulp
- 1¹/₂ teaspoons (4.5 g) mustard powder
- 5 or 6 orange slices, very thinly sliced
- 5 or 6 onion slices, very thinly sliced

Preheat the oven to 450°F (230°C). Remove any pin feathers from the duck and clean it inside and out. Prick the skin so the fat can easily drain from the meat.

In a small bowl, combine the honey, garlic salt, onion salt, poultry seasoning, paprika, and salt. Rub all over the inside and outside of the duck. Place in a shallow baking pan and bake for 15 minutes. Quickly remove, reduce the oven to 350°F (180°C), and drain the fat from the pan. Return the pan to the oven and bake for 1 hour, draining fat as needed. Meanwhile, combine the orange juice and mustard. After 1 hour, remove the duck and brush with some of the orange juice mixture. Using toothpicks, cover the duck with the orange and onion slices. Return to the oven for another 45 minutes, brushing occasionally with the orange juice mixture. Let stand for at least 15 minutes before cutting and serving.

Makes 2 or 3 servings

Mango Chicken

- 4 chicken breast fryer quarters
- 3/4 cup (175 ml) white wine vinegar
- 1 teaspoon fresh thyme leaves
- 1/2 teaspoon salt
- 1/4 cup (60 g) brown sugar
- 1/4 cup (85 g) honey (use a dark and strong summer or fall wildflower—the darker the better)
- 2 teaspoons (4 g) ground ginger
- 2 cloves garlic, minced
- 2 tablespoons (20 g) finely chopped onion
- 2 ripe mangoes, peeled and coarsely chopped

Place the chicken in a large glass bowl. In a measuring cup, mix together 1/2 cup (120 ml) of the vinegar and the thyme and pour over the chicken. Cover and let marinate in the refrigerator for 2 to 3 hours, turning 3 or 4 times.

Preheat the oven to 325°F (170°C). Remove the chicken from the marinade and arrange in a shallow baking dish. Sprinkle with salt. In a small saucepan, combine the brown sugar, honey, ginger, garlic, onion, and the remaining 1/4 cup (55 ml) vinegar. Add the mangoes and bring the mixture to a boil over medium heat. Pour over the chicken, and bake for 50 minutes, or until tender. When done, remove from the oven but keep warm. Drain the juice from the baking dish into a saucepan (or reuse the saucepan you used earlier) and stir over medium heat for 3 to 5 minutes, or until thickened. Pour over the chicken.

Makes 4 or 5 servings

Oven-Fried Chicken in Honey Sauce

- 1 cup (110 g) flour
- 2 teaspoons (12 g) salt
- 1/2 teaspoon freshly ground black pepper
- 2 teaspoons (12 g) paprika (optional)
- 1 cup (225 g) butter
- 3 pounds (approx. 1.5 kg) chicken (a cut-up fryer, or any choice of parts)
- 1/4 cup (85 g) honey (use a strong variety such as goldenrod or buckwheat for a barbecue-like flavour, or a mild variety such as maple or persimmon for a sweet touch)
- 1/4 cup (60 ml) lemon juice

Preheat the oven to 375°F (190°C). In a shallow pan, combine the flour, salt, pepper, and paprika, if using. In a small pan, melt the butter and pour into another shallow dish. Roll the chicken pieces in the melted butter and then roll in the flour mixture to coat. Place the chicken in a baking dish and bake for 30 minutes. Meanwhile, in a small bowl, combine the honey and lemon juice. Remove the chicken from the oven, turn, and pour some of the honey sauce over it. Bake for another 30 minutes, or until tender, basting occasionally with the honey sauce.

Makes 4 or 5 servings

Spaghetti Sauce

- 2 tablespoons (30 ml) extra virgin olive oil
- 1 green bell pepper, finely chopped
- 2 cloves garlic, finely chopped
- 1 large sweet onion, finely chopped
- ¹/₄ to ¹/₂ cup (15 to 30 g) finely chopped fresh parsley
- ¹/₄ cup (30 g) finely chopped celery
- 1 pound (455 g) ground beef
- ¹/₂ pound (228 g) ground pork
- 1 can (16 ounces [475 ml]) tomatoes
- 1 can (15 ounces [440 ml]) tomato paste
- 2 tablespoons (40 g) honey (choose a strong, flavourful variety, such as citrus or alfalfa)
- ¹/₄ teaspoon oregano
- 1 pound (455 g) sliced mushrooms
- **Salt and freshly ground black pepper**

In a large skillet, heat the oil over medium heat. Add the pepper, garlic, onion, parsley, and celery and cook until tender. When tender, add the beef and pork and brown slightly. Pour off excess fat, then add the tomatoes, tomato paste, honey, oregano, mushrooms, and salt and pepper to taste. Combine well. Simmer for 2 hours.

Makes 4 to 6 servings

Sweet Surprise Steak Roll

- 1 pound (455 g) boneless beef top sirloin steak, about ³/₄ inch (about 2 cm) thick
- 2 teaspoons (10 ml) extra virgin olive oil
- 1 teaspoon butter
- 1 teaspoon (3 g) garlic powder
- ¹/₄ teaspoon freshly ground black pepper
- ³/₄ teaspoon salt
- 2 bell peppers (any colors), sliced lengthwise into ¹/₂-inch (1.25-cm) strips
- 1 small white onion, thinly sliced
- 1 cup (70 g) sliced fresh mushrooms
- ¹/₃ cup (40 g) chopped walnuts
- 1 generous tablespoon (15 g) sour cream
- 1 tablespoon (20 g) honey (a mild and sweet variety; clover is best)
- Lemon slices
- Parsley sprigs

Preheat the oven to 250°F (120°C). Pound the steak with the flat side of a meat mallet to about ¹/₄ inch (about 6 mm) thick. Meanwhile, in a large frying pan, heat 1 teaspoon (5 ml) of the oil and the butter on medium-high heat until hot. Sprinkle ¹/₂ teaspoon of the garlic powder and the black pepper over the steak. Fry the steak for 5 to 7 minutes, until done to your liking, turning once. Remove to a heated platter, sprinkle with ¹/₂ teaspoon of the salt, cover with foil, and place in the oven.

Add the remaining 1 teaspoon (5 ml) oil to the frying pan and add the bell peppers, onion, mushrooms, and walnuts. Cook for about 2 minutes, stirring frequently. Add the remaining ¹/₂ teaspoon garlic powder and the remaining ¹/₄ teaspoon salt and cook until the vegetables are tender, stirring frequently.

In a small bowl, mix the sour cream and honey. Remove the steak from the oven and spread with the sour cream mixture. Starting with the short side, roll the steak jelly-roll fashion, holding it in place with 4 or 5 toothpicks along the length of the steak roll. Place the steak on a large platter and add the vegetables. Garnish with lemon slices and parsley sprigs and slice between the toothpicks.

Makes 4 servings

"Oh, My!" Grilled Salmon

- 4 salmon fillets (4 to 6 ounces [115 to 170 g] each)
- ¹/₄ cup (60 ml) peanut oil
- 2 tablespoons (30 ml) soy sauce
- 2 tablespoons (30 ml) balsamic vinegar
- 4 tablespoons (24 g) chopped scallions
- 1 tablespoon (15 g) brown sugar
- 1 tablespoon (20 g) honey (strong and sweet; a citrus honey is an excellent choice)
- 2 cloves garlic, minced
- ¹/₂ teaspoon sesame oil
- ¹/₄ teaspoon salt

Place the fillets in a glass dish. In a medium bowl, blend together the peanut oil, soy sauce, vinegar, scallions, brown sugar, honey, garlic, sesame oil, and salt, and pour over the salmon. Cover the bowl tightly and marinate in the refrigerator for at least 4 hours, 6 is better. Save the marinade. Prepare the grill so the heat is about medium to medium-high. Oil the grill well and place the fillets on the grill for about 10 minutes per inch (cm) of thickness, at the thickest part. Turn halfway through cooking; salmon is done when it just barely flakes when tested with a fork.

If desired, add 1 tablespoon (10 g) of honey and 1 tablespoon (15 ml) of balsamic vinegar to the reserved marinade. Place in a saucepan and bring just to a boil over medium heat, remove, and brush on the salmon just before serving.

Makes 4 servings

Spring Lamb Roast

This is a perfect recipe for honeys infused with thyme, rosemary, or otherwise (see page 120).

ROAST

- 2 tablespoons (40 g) honey (use a flavourful variety such as basswood or a summer wildflower)
- 1 tablespoon (15 g) French mustard
- 5-pound (approx. 2.5 kg) leg of lamb
- Salt and freshly ground black pepper
- ¹/₄ cup (10 g) thyme sprigs, slightly crushed
- ¹/₄ cup (8 g) rosemary leaves, slightly crushed

GRAVY

- 1 tablespoon (15 g) flour
- 2 to 3 jiggers good brandy
- 1 cup (235 ml) beef stock (or lamb stock, if you can find it)
- 1 tablespoon (20 g) honey (use same as above)

For the roast: Preheat the oven to 350°F (180°C). In a small bowl, combine the honey and mustard and mix well. Rub all over the roast, using your hands, not a brush. Season with the salt and pepper to taste. Put the roast on a wire rack that is sitting in a roasting pan. Bake for 1 hour and 40 minutes. Before the final 30 minutes of baking, sprinkle the thyme and rosemary leaves all over the roast and then cover with foil so it doesn't become overly browned. Remove from the oven and let rest for at least 15 minutes before serving.

For the gravy: Pour the drippings from the roast into a medium saucepan, add the flour, and stir over low heat for a minute or two. Pour in the brandy to taste and stir to combine. Gradually add the stock, continuing to stir. Bring just to a boil, turn down the heat to simmer, and continue to stir for another 2 to 3 minutes, adding the honey at the last minute so as much of the flavour and aroma remains as possible.

Makes 8 to 10 servings

Desserts

Wild Blueberry Pie

- 6 cups (870 g) fresh wild blueberries, divided
- ³/₄ cup (255 g) honey (medium to light flavoured, such as a summer wildflower or blueberry)
- ¹/₄ teaspoon salt
- ¹/₄ cup (60 ml) water
- 4 tablespoons (32 g) cornstarch
- 2 teaspoons butter
- 2 teaspoons (10 ml) lemon juice
- 1 prebaked 9-inch (22.5-cm) pie crust (or a homemade baked and cooled pie crust)
- Whipped cream (optional)

In a large saucepan, combine 5 cups (725 g) of the blueberries with the honey and salt. In a glass, mix the water and cornstarch and add to the blueberries. Cook over medium heat until the mixture bubbles. Then add the butter, lemon juice, and the remaining 1 cup (145 g) of blueberries and stir. Remove from the heat and let cool just a bit, then pour into the pie crust. Chill. When serving, top with whipped cream, if desired.

Makes 6 servings

Pecan Pie

- ¹/₂ cup (170 g) honey (use the lightest, mildest variety you can find)
- ¹/₂ cup (115 g) brown sugar
- ¹/₄ cup (55 g) butter
- 3 eggs, beaten
- 1 ¹/₂ cups (165 g) broken pecan meats
- ¹/₂ teaspoon cinnamon
- ¹/₂ teaspoon vanilla extract
- 9-inch (22.5-cm) pie pan, lined with ready-to-bake pie crust

Preheat the oven to 400°F (200°C). In a medium saucepan, blend the honey and brown sugar together over medium heat. Cook slowly until they form a smooth syrup. Add the butter, blend in carefully until melted. Add the eggs, cinnamon, vanilla extract, and the pecans and mix well, coating the nuts. Pour the mixture into the pie pan. Bake for 10 minutes, then reduce the oven temperature to 350°F (180°C) and bake for an additional 25 to 30 minutes. Remove from the oven and let cool. Cut only when firm.

Makes 6 servings

Old-Fashioned Bread Pudding

For the bread cubes, use a potato bread, an egg bread, or some other bread that's not too strongly flavored. Either dark or golden raisins will be delicious in this bread pudding—choose whichever is your favorite.

- 8 cups (400 g) bread cubes (about ³/₄ inch [2 cm] cubes)
- 1 cup (235 ml) half-and-half
- ¹/₂ cup (170 g) honey (choose a variety that's strong, dark, and flavourful, such as a late-summer wildflower)
- 1 teaspoon (5 ml) vanilla (or a bit more to taste)
- 1 heaping teaspoon (25 g) ground cinnamon
- 3 cups (710 ml) whole milk
- 6 eggs
- 1 tablespoon (6 g) grated orange zest
- ¹/₂ cup (75 g) raisins
- Honey Whipped Cream (see recipe below)

Place the bread cubes in the bottom of a barely greased shallow 2-quart baking pan. In a large bowl, thoroughly combine the half-and-half, honey, vanilla, cinnamon, milk, eggs, and orange zest. Fold in the raisins. Pour over the bread cubes and let stand for at least an hour, until all the liquid is absorbed by the cubes. Meanwhile, preheat the oven to 375°F (190°C). Bake for 45 minutes, or until a knife or toothpick inserted into the center comes out clean. Top with Honey Whipped Cream when serving.

Honey Whipped Cream: Beat 1 cup (235 ml) whipping cream until fluffy. Slowly add ¹/₄ cup (85 g) of the lightest-flavored honey you can find, continue beating, and then add 1 teaspoon (5 ml) vanilla and beat until stiff peaks form.

Makes 10 to 12 servings

Honey Fudge Pudding Cake

If you're so inclined, you can certainly substitute one of your own homemade chocolate cake recipes for the prepared cake mix.

- 1 package (18.25 ounces [515 g]) chocolate cake mix
- 1 cup (125 g) coarsely chopped pecans or walnuts
- ¹/₂ cup (112 g) butter
- ¹/₂ cup (120 ml) water
- 1 cup (340 g) honey (use a medium summer wildflower, citrus, alfalfa, or starthistle)
- 6 to 8 ounces (170 to 225 g) semisweet chocolate morsels

Heat the oven to 350°F (180°C), or the temperature recommended for your specific cake mix. Grease a 13 x 9 x 2-inch (32.5 x 22.5 x 5-cm) cake pan with vegetable shortening. Following the recipe on the box, mix the cake ingredients in a large mixing bowl and add the nuts. Pour into the cake pan. In a small saucepan, combine the butter, water, and honey and heat to boiling. Scatter the chocolate morsels (to taste) over the top of the cake mix, and then pour the honey mixture over all. Bake for 45 to 50 minutes (add about 5 minutes if setting the time by the box mix to accommodate the extra liquid). Test with a toothpick. Serve warm with French vanilla ice cream.

Makes 15 to 20 servings

Feed Your Sweet Tooth

In this section, I've tried to come up with "normal" serving sizes. But when it comes to feeding your sweet tooth, sometimes rules are made to be broken! So, depending on how you plan on serving them, the recipe yields may change.

Almond Tarts

Shells made this way will be a tad uneven because they are so small. To grind the almonds, use a clean coffee grinder or food processor, and grind them with the sugar to a fine consistency.

- 1/3 cup (75 g) butter
- 1/3 cup (66 g) sugar
- 1 teaspoon honey (try a light, mild early spring variety such as black locust or even maple)
- 1/3 cup (42 g) ground raw almonds
- 4 cups (520 g) fresh raspberries, blackberries, or black raspberries (mix, match, or keep to a single variety)
- 1 pint (475 ml) whipping cream, whipped with sugar to taste

Preheat the oven to 350°F (180°C). Spray a mini-muffin tin or miniature tart pans with cooking spray. In a medium bowl, using an electric mixer, cream together the butter, sugar, honey, almond extract, and almonds.

Spread about 1 teaspoon of the creamed mixture along the bottom and up the sides of each muffin tin or tart pan. Be sure to cover the bottom of each cup completely and spread as far up the sides as possible. Try to keep the tops even. Bake for 10 minutes, or until golden brown. The tarts will be bubbly, but will set upon cooling.

Allow the shells to cool only partially before removing them from the containers. They will become brittle if left in the containers too long. You can mix your berries and whipped cream together before filling, or fill with berries and top with the whipped cream, but do so just before serving to avoid having the shells become soggy.

Makes 1 1/2 dozen mini tarts

Orange Cake

- 2 cups (220 g) cake flour
- 3/4 teaspoon salt (optional)
- 3/4 teaspoon baking soda
- 1/2 cup (112 g) butter
- 1 cup (340 g) citrus honey, or, even better, starthistle or even black locust honey
- 2 eggs
- 1/4 cup (60 ml) condensed milk
- 2 tablespoons (30 ml) freshly squeezed orange juice
- 1 1/2 teaspoons (3 g) finely grated orange zest

Preheat the oven to 350°F (180°C). Grease a 9-inch (22.5 cm) square pan. In a medium bowl, sift together the flour, salt, and baking soda and set aside. In a large bowl, cream the butter slightly with an electric mixer. As you cream the butter, add the honey in a fine stream. Add the eggs one at a time, mixing well after each. In another bowl, combine the condensed milk, orange juice, and orange zest.

Add about a quarter of the flour mixture to the butter mixture and mix well, then about a quarter of the milk mixture, beating continuously. Continue alternating adding dry and wet ingredients.

Pour into the pan. Bake for 45 to 50 minutes. Cool before serving.

Makes 9 servings

Regular Ol' Fudge

- 2 ounces (55 g) unsweetened chocolate
- ¹/₄ teaspoon salt
- ²/₃ cup (160 ml) milk
- 2 tablespoons (30 ml) corn syrup
- ¹/₂ cup (170 g) honey (use a medium- to light-flavoured variety; clover is great in this recipe)
- 2 cups (400 g) sugar
- 2 tablespoons (28 g) butter
- 1¹/₂ teaspoons (7.5 ml) vanilla
- ¹/₂ cup (63 g) chopped walnuts or pecans

Butter a 8 x 8-inch (20.3 x 20.3 cm) pan liberally, then set aside. In a 2-quart (1.9 L) saucepan, combine the chocolate, salt, milk, corn syrup, honey, and sugar. Cook over medium heat, stirring continuously, until the chocolate is melted and the honey is dissolved. Using a candy thermometer, continue cooking and stirring until the mixture reaches 236°F (113°C), or the soft ball stage. Remove from the heat, add the butter, and set aside to cool until the mixture reaches 120°F (49°C). Add the vanilla and beat vigorously until the mixture is thick and loses its gloss. Stir in the nuts and spread in the pan. Don't cut until the fudge has set up and cooled. If you can wait, it actually tastes better the next day.

Makes 20 servings

Banana Popsicles

Figure on a generous ¹/₂ cup (63 g) of peanuts per banana.

- Fresh, firm bananas
- Popsicle sticks
- Honey (use a strong, flavourful variety)
- Chopped, salted peanuts

Peel the bananas and cut them in half crosswise. Insert a popsicle stick in the cut end of each banana half. Dip each banana half in the honey all the way to the stick. Roll in the peanuts until completely covered. Place in a nonstick pan or on a small baking sheet lined with wax paper and freeze until solid. (Or, if you're able, stand them up in your freezer.) Once frozen, store each in a sealable sandwich bag.

Baklava, the Honey Dessert King! ☞

HONEY SYRUP

- 2 cups (475 ml) warm water
- 2²/₃ cups (535 g) sugar
- 4 whole cloves
- 4 cinnamon sticks
- Rind of a small lemon, cut into strips
- 2 cups (455 g) honey (use a rich, full-bodied dark variety, such as sumac or tulip poplar, or use a honey infused with thyme [see the recipe for making infused honey on page 120])
- 5 tablespoons (75 ml) fresh lemon juice

For the syrup: In a medium saucepan, combine the water, sugar, cloves, cinnamon, and lemon rind and bring to a boil. Reduce the heat and cook for 10 to 15 minutes, until the liquid begins to thicken. Stir in the honey and lemon juice. Stir for a few minutes, until the honey dissolves and the lemon juice begins to fume. Remove from the heat and strain out the rind and spices. Keep in a jar at room temperature. Label the jar Honey Syrup.

PASTRY

- 1 package (16 ounces [455 g]) fresh phyllo pastry sheets
- ¹/₂ cup (112 g) butter, melted
- 1¹/₂ cups (190 g) finely chopped walnuts
- 1¹/₄ cups (295 ml) Honey Syrup

For the pastry: Preheat the oven to 350°F (180°C). Unroll the phyllo dough. Separate out a stack of 5 to 8 sheets, and place an 8-inch (20-cm) round cake pan on top of the stack close to all the edges. Cut around the cake pan with a sharp knife to make circles. Repeat the process until all of the phyllo dough is used.

Butter the cake pan liberally. Place one circle of dough on the bottom of the pan and brush with the melted butter. Repeat for 5 layers of dough. Sprinkle ¹/₄ cup (32 g) of the walnuts over this last, buttered layer. Cover the walnuts with a single layer of dough, brush with butter, and put 2 tablespoons of walnuts on the buttered layer. Repeat for the next 15 to 18 layers, depending on the walnuts and dough. As you build the stack, on occasion gently press down to make the layers fit in the pan. Finish with 5 buttered layers of phyllo dough. With a very sharp knife, mark in pie-shaped pieces ¹/₂ inch (1.3 cm) deep or so. Bake for 45 minutes, or until golden brown. Remove from the oven, and while still hot, pour the Honey Syrup over the pastry. Cut the marked pieces all the way through and let stand for a day before eating. If you can wait that long.

Makes 8 servings

Strawberry Shortcake ☞

- **2 cups (280 g) fresh strawberries, halved, or quartered if very large**
- **¹/₄ cup (60 ml) balsamic vinegar**
- **¹/₄ cup (80 g) honey (use a medium-strong variety such as summer wildflower or clover)**
- **1 tablespoon (15 g) brown sugar**
- **¹/₄ cup (60 ml) whipping cream**
- **1 teaspoon (5 ml) vanilla**
- **6 shortcakes (ready-made sponge cakes, homemade biscuits, etc.)**
- **Sprigs of mint, for garnish**

In a medium mixing bowl, combine the strawberries, vinegar, 2 tablespoons (40 g) of the honey, and the brown sugar. Set aside. In a chilled mixing bowl, combine the whipping cream and vanilla. With an electric mixer, whip until soft peaks form. Stream in the remaining 2 tablespoons (40 g) honey and continue beating until stiff peaks form.

To serve, divide shortcakes in half for a top and bottom, place the bottom on a dessert plate, and spoon on the strawberries. Top with some whipped cream, cover with the top of the shortcake, add additional whipped cream, and garnish with mint.

Makes 4 to 6 servings

Strawberries & Cream

- **2 cups (280 g) fresh strawberries, halved, or quartered if large**
- **1 tablespoon (15 ml) vanilla**
- **¹/₄ cup (50 g) sugar (or to taste)**
- **¹/₂ cup (170 g) honey (the best for this is starthistle, but clover or locust is an okay second)**

In a medium bowl, combine the strawberries, vanilla, sugar, and honey. Using a potato masher, mash the berries until juicy and generally pulverized. Let sit in the fridge overnight and serve over vanilla ice cream.

Makes 2 to 3 servings

Pastries and Baked Goods

Plain Jane White Bread

The first time you make this recipe, I recommend using a light, mild variety of honey. But the next time you make it, try something else to see what a difference it makes. Just don't go too dark, because it burns easily. This recipe makes two loaves, so freeze one for later.

- 2 cups (475 ml) warm water
- 2 envelopes (2¼ teaspoons [9 g] each) dry yeast
- ½ cup (170 g) honey
- 1 teaspoon (6 g) salt
- ¼ cup (50 g) vegetable shortening
- 1 egg
- 7 cups (770 g) flour

Pour the water into a large mixing bowl. Stir in the yeast until dissolved. Stir in the honey, salt, shortening, and egg. Mix in enough flour to make the dough easy to handle. Knead for 5 to 6 minutes. Sprinkle just a little flour on the ball of dough so it's covered, return the dough to the bowl, cover with a damp towel, and let rise at room temperature for 1½ to 2 hours. Once raised, knead for a few minutes until reduced to about the size it was before. Divide the dough into two pieces and form each into a log. Fit each log into a greased 9 x 5-inch (22.5 x 13 cm) loaf pan. Let rise again for an hour or so, while heating your oven to 400°F (200°C). When the dough rises half again as tall as the pan, place in the oven and bake for 20 minutes. Reduce the heat to 350°F (180°C) and bake for another 20 minutes. Watch so the crust doesn't get too brown because of the honey. Remove from the oven, turn out onto a rack, and let cool.

Makes 2 loaves

Sweet Fall Bread

- 1 teaspoon (4.5 g) baking soda
- 1 cup (235 ml) buttermilk
- 1 cup (230 g) sour cream
- 2 tablespoons (40 g) honey (a medium fall variety such as goldenrod mixed with aster, or golden honey plant)
- 2 eggs
- ¼ cup (62 g) mashed cooked fall squash, such as acorn or butternut
- 2 cups (275 g) cornmeal
- 1 teaspoon (6 g) salt
- 1½ teaspoons butter

Preheat the oven to 350°F (180°C). In a small bowl, stir the baking soda into the buttermilk until it dissolves. Stir in the sour cream, and then slowly add the honey. In a medium bowl, beat the eggs and add the squash, cornmeal, salt, and the buttermilk mixture. Put the butter in a 9-inch (22.5-cm) baking pan and place in the oven until melted. Coat the pan, sides and bottom, with the butter, then add the remaining butter to the batter. Fill the pan and bake for 35 to 40 minutes.

Makes 6 to 8 servings

Sweet Zucchini Bread ☞

Sure, you can make this zucchini bread without nuts, but pecans are hard to beat in this recipe!

- **3 eggs**
- **1 cup (235 ml) vegetable oil**
- **1 cup (340 g) honey (use a medium- to light-flavoured honey, so it adds a hint of flavour, maybe one infused with lemon or garlic)**
- **2 cups (280 g) grated zucchini, unpeeled for color**
- **1 tablespoon (15 ml) vanilla**
- **3 cups (330 g) flour**
- **1 teaspoon (4.5 g) baking soda**
- **¹/₂ teaspoon baking powder**
- **1 teaspoon (6 g) salt**
- **1 cup (125 g) finely chopped nuts (optional)**
- **¹/₂ teaspoon cinnamon (optional, but shouldn't be)**

Preheat the oven to 325°F (170°C). In a large mixing bowl, beat the eggs. Mix in the oil, honey, zucchini, and vanilla until combined. Using another large mixing bowl, combine the flour, baking soda, baking powder, and salt. Add the egg mixture and stir only enough to moisten the dry ingredients. Add the nuts and cinnamon, if using, and stir just enough to combine. Divide into 2 well-greased bread pans. If using more than 2 pans, watch your time, but bake for about 1 hour. Smaller pans may finish sooner. Test with a toothpick.

Makes 2 loaves

Just Barely Sweet Biscuits

- 2 cups (220 g) flour
- 2 teaspoons (9 g) baking powder
- ¹/₂ teaspoon salt
- ¹/₂ cup (112 g) cold butter
- ¹/₄ cup (85 g) light, very mild honey
- 1 tablespoon (15 ml) whole milk

Preheat the oven to 425°F (220°C). Sift the flour, baking powder, and salt together in a large bowl. In a small bowl, blend the butter, honey, and milk together. Then slowly add the milk mixture to the dry ingredients (it's okay to use a little more milk if necessary, but not too much, so the dough just barely holds together). Pour the dough out onto a pastry board or cutting board, and form it into a 6 x 9-inch (15 x 22.5-cm) rectangle. Cut this into 6 squares, and then cut the squares diagonally into 2 triangles. Carefully lift onto a baking sheet, spreading them out enough so they don't touch. Bake for 10 to 12 minutes (depending on how much milk you used), or until the tops turn a lovely shade of brown. Remove and serve warm with butter and—what else? —honey.

Makes 6 servings

Orange Cranberry Rolls

- 2 cups (220 g) bread flour
- ¹/₂ cup (40 g) old-fashioned oats
- 1¹/₂ teaspoons instant yeast (also known as rapid rise)
- ¹/₂ teaspoon salt
- ¹/₂ cup (75 g) dried cranberries, coarsely chopped
- 1 large egg
- ¹/₄ cup (60 ml) milk
- 1 tablespoon (6 g) orange zest
- 1 teaspoon (5 ml) vanilla
- ¹/₃ cup (80 ml) freshly squeezed orange juice, at room temperature
- ¹/₄ cup (85 g) honey, at room temperature (use a medium variety, such as alfalfa, citrus, or a midsummer wildflower)
- 1 tablespoon (14 g) butter, softened

GLAZE

- 2 tablespoons (30 ml) orange juice
- 1 tablespoon (14 g) butter, melted
- Coarse sugar, for sprinkling (optional)

You'll need three mixing bowls to begin with—one large, one small, and one microwavable. (If you plan on using a stand mixer to knead your dough, use the stand mixer's bowl as your large bowl.)

In the large bowl, combine the flour, oats, yeast, salt, and cranberries. In the small bowl, beat the egg with the milk, orange zest, and vanilla. Set both bowls aside.

In the microwavable bowl, combine the orange juice, honey, and butter. Heat the mixture in the microwave until the butter is melted and the liquid is very warm, but not hot.

Add the milk mixture and the orange juice mixture to the dry ingredients and mix well to incorporate. If using a stand mixer, use the dough hook and run the mixer for 10 to 15 minutes to knead the dough. If kneading by hand, place the dough on a floured surface and knead for at least 15 minutes, until the dough is soft and supple. To test if you've kneaded the dough enough, rip off a golf ball–size piece of dough and roll it into a smooth ball. Working gently, use both hands to stretch the dough apart. If the dough stretches without ripping to create an almost see-through "window," it's ready. If it rips, continue kneading a bit longer.

Place the dough in a greased bowl, turn once to coat, and cover loosely. Let rise in a warm place for about 1¹/₂ hours, or until the dough is puffy and has nearly doubled in size. (An oven set to "warm" for a few minutes and then switched off works well for this, as does placing the bowl on a rack over a pan of hot water.)

Punch down the dough and shape into 12 rolls. Place the rolls in a greased cupcake tin, cover, and let rise in a warm place for about 45 minutes. Preheat the oven to 375°F (190°C) as your rolls finish their final rise.

For the glaze: Mix the orange juice and butter together and brush onto the tops of the rolls. Sprinkle with sugar and bake for about 15 minutes, or until the tops of the buns are light golden in colour.

Serve warm with butter—and honey, if you like!

Makes 12 rolls, to die for

Blue Honey Muffins

- ¹/₂ cup (120 ml) orange juice, with pulp
- ¹/₃ cup (75 g) butter, melted
- ¹/₂ teaspoon vanilla
- ¹/₂ cup (170 g) honey (use the one you like the best, but stronger honeys will have more flavour in the final product)
- 2 eggs, well beaten, but not frothy
- 2 cups (220 g) flour
- ¹/₄ cup (50 g) sugar
- 2 teaspoons (9 g) baking powder
- 1 teaspoon (4.5 g) baking soda
- ¹/₂ teaspoon salt
- 1 cup (145 g) fresh blueberries (wild blueberries are the best)
- (Optional: coarse sugar for topping and one insulated carafe good coffee)

Preheat the oven to 375°F (190°C). Grease and flour a muffin tin.

In a medium bowl, combine the orange juice, butter, vanilla, honey, and eggs and mix well. In a large bowl, combine the flour, sugar, baking powder, baking soda, and salt. Slowly add the liquid mixture to the dry mixture, mixing only until blended. Add the blueberries, stirring just enough to equally disperse the berries throughout the batter. Pour the batter into the muffin cups, filling each about two-thirds full. Bake for 15 to 20 minutes, until golden and a toothpick inserted into the center comes out clean. Let cool.

Take 5 or 6 with you to the beeyard when you harvest honey. Halfway done, stop for a break with a muffin and a cup of that good coffee.

Makes 12 muffins

Cookie-Making Hint

The bigger you make the Honey Pot Cookies, the bigger the indent can be, and the more filling you can add.

Honey Pot Cookies

- ¹/₂ cup (112 g) butter
- ¹/₃ cup (115 g) honey (use a mild, not overpowering variety from a spring honey flow)
- 1 egg
- ¹/₂ teaspoon vanilla
- ¹/₄ teaspoon salt
- 1 cup (110 g) flour
- Any fruit jam, jelly, or other filling

Preheat the oven to 375°F (190°C). In a medium bowl, blend together the butter, honey, egg, and vanilla. Stir in the salt. Add the flour and stir until combined. When sticky, roll into balls about 1 inch (2.5 cm) in diameter. Place on a baking sheet and bake for just 5 minutes. Remove, and using the tip of your finger or a similarly sized object, indent the balls to about the center, leaving a fairly large hole. Quickly return to the oven and finish baking for another 7 to 10 minutes. Allow to cool. Fill with your choice of jams, jellies, sauces, creams, cheese spreads—anything is good in these.

Makes 12 cookies

Sweet Chocolate Brownies

- **2 ounces (55 g) unsweetened chocolate**
- **¹/₃ cup (75 g) butter**
- **³/₄ cup (83 g) flour**
- **³/₄ teaspoon baking powder**
- **¹/₄ teaspoon salt**
- **2 eggs**
- **³/₄ cup (255 g) honey (use a medium flavourful wildflower variety from late spring or early summer)**
- **¹/₂ cup (63 g) finely chopped pecans**
- **¹/₂ cup (63 g) coarsely chopped walnuts**

Preheat the oven to 325°F (170°C). In a small saucepan, place the chocolate and butter and warm over low heat until melted. In a medium bowl, combine the flour, baking powder, and salt. Set aside. In a large bowl, beat the eggs with an electric mixer until thick and fluffy. Very gradually add the honey in a fine stream. Gradually add the chocolate mixture. Then add the egg-and-chocolate mixture to the dry ingredients, mixing constantly on low speed. Blend in the pecans, followed by the walnuts. When mixed, pour into a greased 8-inch (20-cm) square baking dish and bake for 25 to 30 minutes, until a toothpick inserted into the center comes out clean.

Makes 16 servings

Short and Sweet

Banana Morsels

For the best flavour, combine several varieties of nuts. Pecans, walnuts, and peanuts work particularly well. The beauty of this recipe is that it can easily be adjusted depending on the number of hungry people in your house looking for a snack. Simply count on a banana per person, and figure a generous ½ cup (63 g) of nuts per banana.

- **Coarsely ground nuts (use several varieties)**
- **Bananas, peeled and cut into bite-size pieces**
- **Honey (use a medium to dark full-flavoured variety)**

Place the nuts in a flat pan. Using toothpicks or hors d'oeuvres spears, dip the banana pieces into the honey, completely submerging them. Allow to drain for a moment, then roll in nuts so they are completely covered. Chill. Serve.

Makes 4 to 6 pieces per banana

Berry Crunch

- **3½ cups (455 g) fresh blackberries or raspberries (use whole if you don't mind the seeds, or crush and strain to remove seeds)**
- **1 cup (340 g) honey (berry honey is the obvious choice, though black locust or any light, mild spring honey works well)**
- **1 cup (110 g) flour**
- **½ cup (40 g) oats**
- **½ cup (115 g) brown sugar**
- **½ cup (36 g) shredded coconut**

Preheat the oven to 350°F (180°C). Grease a 9-inch (22.5 cm) pie plate. In a large bowl, mix the berries, honey, and flour together. Pour into the pie plate. Sprinkle with the oats, brown sugar, and coconut. Bake for 45 to 50 minutes, or until the surface is richly browned.

Makes 6 servings

Sweet Cheese for Breakfast Spread

- **1 cup (250 g) ricotta cheese**
- **½ cup (115 g) unflavored yogurt**
- **3 to 4 teaspoons (20 to 27 g) honey (use a medium- to strong-flavoured variety, so you can taste it over the yogurt; a summer or fall wildflower blend works well here)**
- **½ cup (120 ml) whipping cream**

In a medium bowl, crumble the ricotta cheese, then add the yogurt and honey and beat well. In another bowl, beat the whipping cream just until soft peaks form. Fold the cheese mixture into the whipped cream. Place in a serving bowl. This is excellent on croissants or specialty breads, or as a dip for fresh, seasonal fruit with a cup of strong coffee.

Makes 4 servings (coffee, croissants, and Sunday newspaper optional)

Sweet Cheese Anytime Spread

- **8 ounces (230 g) cream cheese, at room temperature**
- **2 tablespoons (40 g) honey (use a medium-flavoured variety, such as citrus or a summer wildflower)**
- **1 to 2 teaspoons (2 to 4 g) finely grated orange or lemon zest**

In a medium bowl, combine the cream cheese, honey, and zest to taste. Beat until well mixed and fluffy. This is fantastic on brownies, croissants, muffins, cakes, and other pastries.

Makes 1 cup (230 g)

Honey Blue Heaven Spread 🐝

For the best flavor, get wild blueberries from Maine or the Maritime Provinces, if you can.

- ¹/₂ cup (73 g) wild, fresh blueberries
- ¹/₄ cup (80 g) honey (very mild, or try blueberry honey, which isn't quite so mild)
- ¹/₂ cup (112 g) butter, at room temperature

In a small saucepan, combine the blueberries and 2 tablespoons (40 g) of the honey, stirring just enough to cover the berries with honey. Slowly bring to a boil over medium-high heat, stirring constantly. Once boiling, continue stirring until the mixture is thick and reduced to about half of what it was. Pour in the remaining 2 tablespoons (40 g) honey, and blend in the butter until thoroughly combined. Remove from the heat. Keep refrigerated. Use on croissants, rolls, scones, or crêpes.

Makes 2 or 3 servings

QUICK TASTE

Orange Honey Butter

Into the bowl of an electric mixer, place ¹/₂ cup (112 g) of room-temperature butter with ¹/₃ cup (115 g) of a moderately flavoured, medium-coloured honey (ideally, orange blossom honey). A light honey will be sweet, but offers little flavour. Cream the butter and honey together, then add 1 teaspoon (2 g) of finely grated orange zest and finish creaming the mixture. Whip this up fresh just before you are going to use it—it is absolutely perfect on croissants at breakfast.

Makes 1 cup (230 g)

Creamy Fruit Dressing

*Don't even think about using anything but real mayonnaise
for this recipe!*

- 1 cup (240 g) mayonnaise
- ¹/₄ cup (85 g) honey (a summer wildflower or even a fall
  wildflower works well, something with a bit of flavour)
- 1 tablespoon (6 g) grated orange zest
- 2 tablespoons (30 ml) freshly squeezed orange juice

In a small bowl, combine all the ingredients, mixing well, and
then chill. Use as a sauce over fruit, or as a dip.

Makes 1 ¹/₂ cups (360 g)

Great Granola

- ²/₃ cup (230 g) honey (use a mild but not too mild variety so
  the flavour comes through)
- ³/₄ cup (195 g) extra-chunky peanut butter
- 4 cups (488 g) granola mix

Pour the honey into a 4-cup (950 ml) microwavable container
and heat on high for 2 to 3 minutes, until the honey just begins
to boil. Immediately remove from the microwave. Add the
peanut butter and mix until it is thoroughly blended with the
honey. Put the granola in a larger mixing bowl, then pour the
honey/peanut butter mixture over the granola and combine until
the granola is completely covered. Press into a 13 x 9 x 2-inch
(32.5 x 22.5 x 5 cm) baking pan and let sit at room temperature
until firm. Don't refrigerate. When it's firm, cut into squares, and
take two or three to the beeyard for a quick pick-me-up when
hefting heavy honey supers.

Makes eighteen 2 x 3-inch (5 x 7.5 cm) squares

Sweet Mandarin Slices

Remove the peel and pith from a fresh, seedless Mandarin orange and set aside the slices. Heat a skillet on medium, and add $^1/_4$ cup (85 g) of honey—an early spring variety that is light and mild, but as sweet as you can find—with $^1/_4$ teaspoon of fresh grated nutmeg and 2 or 3 tablespoons (28 or 42 g) of butter. (For variety, add 1 tablespoon [6 g] of grated Mandarin zest.) When the butter is melted, add the Mandarin slices to the skillet. Simmer the slices for 10 to 15 minutes, turning them frequently so both sides are glazed evenly. Remove the slices from the skillet when they become soft and translucent. Serve them warm as a side dish with roast chicken, duck, goose, or pheasant, or as a treat on vanilla ice cream.

Makes 1 serving per orange

Apples 'n' Honey

Choose a tart, crisp variety of apple for this recipe.

- 6 cups (900 g) peeled and sliced apples
- $^1/_2$ cup (170 g) honey (apple honey if you have it, but a good substitute is clover or a mild, spring wildflower)
- 1 tablespoon (15 ml) lemon juice
- 2 to 4 tablespoons (28 to 56 g) butter, to taste
- $^1/_2$ to 1 teaspoon (1.3 to 2.5 g) cinnamon, to taste

Preheat the oven to 350°F (180°C). Liberally butter a medium baking dish and assemble the apple slices as evenly as possible in the pan. In a small saucepan, mix together the honey and lemon juice and warm slightly over low heat so the mixture pours easily. Pour over the apples, covering as much as possible. Dot with the butter and sprinkle with cinnamon. Cover with foil and bake for about 20 minutes, or until the apples are tender. As the apples bake, tip the dish to gather the liquid at the bottom and baste occasionally. This is great along with or over ice cream, or with whipped cream or clotted cream on top.

Makes 6 to 8 servings

Sweet Bruschetta

Choose your favorite honey for this unique bruschetta. Whether you use a light, mild variety or a stronger, darker variety, each will add its own distinctive personality to the finished delicacy.

- $^1/_4$ cup (60 ml) extra virgin olive oil
- 3 or 4 cloves garlic, minced
- 1 French baguette, sliced on the diagonal into half-inch (or centimeter) -thick slices
- 1 pound (455 g) Gorgonzola cheese or a mild goat cheese, very thinly sliced
- 1 cup (340 g) honey

In a small bowl or cup, mix the olive oil and garlic together and spoon or brush on one side of each slice of bread. Place these slices, garlic side up, on a cookie sheet under a medium broiler until the garlic begins to turn slightly brown. Don't overcook. Remove from the oven and put a slice or two of cheese on top of each bread slice and return to the broiler just long enough to have the cheese soften, but not melt. Remove from the oven and immediately drizzle your honey over the top. If you'd like, use two or more honeys for variety. The honey should puddle on the cheese, and some should run off just a little for effect. Serve while still warm and the cheese still soft.

Makes 10 appetizer servings

Fruit Smoothie

- $^1/_2$ cup (120 ml) freshly squeezed orange juice, with all the pulp you can manage
- 1 banana, sliced
- $^1/_2$ cup (115 g) plain yogurt
- 2 tablespoons (40 g) honey (use a summer wildflower with good flavour but light colour)
- $^1/_4$ cup (28 g) strawberries
- $^1/_4$ cup (28 g) raspberries
- $^1/_4$ cup (37 g) blueberries

Place all ingredients in a blender and blend until smooth. Enjoy

Makes 1 serving

Honey Lemonade Concentrate and Finisher

TO MAKE THE CONCENTRATE:

- 6 tablespoons (120 g) of an early season wildflower, locust, or mesquite honey
- 1 cup (237 ml) freshly squeezed lemon juice (from about 3–4 lemons)
- 1 whole lemon, very thinly sliced into transparent rounds, seeds removed

In a quart jar, place the honey and lemon juice and shake vigorously to dissolve the honey. (You can also put the honey and lemon in a bowl and whisk together.) When the honey is completely dissolved, add the lemon slices and place in the refrigerator until needed. Chill for at least an hour, but overnight will better allow the flavours to blend.

TO MAKE THE LEMONADE FINISHER:

Fill 12-ounce (355 ml) beverage glasses with ice cubes all the way to the top. Pour in $^1/_4$ cup (60 ml) of the concentrate and fill the glass with carbonated water, tonic water, or just add cold water. The result is wonderfully refreshing after working bees on a hot, hot day. Drink one to cool off, then another to enjoy.

Makes 5 to 6 servings

Mint Tea Punch

[recipe ingredients]1 cup (145 g) fresh blueberries, raspberries, blackberries, and/or strawberries

- 1 cup (340 g) honey (choose a light, mild, but sweet variety)
- 1 $^1/_2$ cups (355 ml) orange juice
- $^1/_2$ cup (120 ml) lemon juice
- 1 quart (946 ml) mint tea

In a large pitcher, crush the berries and add the honey, orange juice, lemon juice, and tea. Combine thoroughly. Chill. Serve over crushed ice. Delightful on a hot, humid afternoon.

Makes 4 servings

ONE COMPLETE MEAL

....

Michael Young is a beekeeper, a meadmaker, and an accomplished chef. He teaches at the Culinary Institute in Belfast, Northern Ireland, and he teaches beekeeping, meadmaking, and encaustic painting (using beeswax) throughout the United Kingdom and the United States.

His unique culinary experiences have permitted him to experiment with cooking with honey, but with an exotic touch. Here are some of the recipes I have borrowed from his kitchen that use honey and a host of other interesting flavours, textures, and aromas.

Beekeeper's Mulled Wine (Wassail)

A comforting drink for a cold evening, a perfect antidote for the stresses and strains of everyday life, and the perfect way to welcome friends into the home.

- 1 quart (0.94 L) bottle mead
- 1 quart (0.94 L) bottle red wine
- 1 cup (235 ml) sweet red vermouth
- 1 cup (235 ml) cranberry juice
- 6 strips orange rind (about half an orange)
- 8 whole cloves
- 1 stick cinnamon
- 8 pods cardamom, crushed
- $^1/_2$ cup (75 g) dark raisins
- $^1/_2$ cup (170 g) honey
- Lemon slices
- Orange slices
- Apple slices

Pour the mead and red wine into a large stainless steel or enamel saucepan. Add the vermouth, cranberry juice, orange rind, cloves, cinnamon, and cardamom pods. Heat the wine mixture gently until very hot, but do not boil. Remove the saucepan from the heat, cover with a lid, and cool. When cool, strain the wine through a colander into a serving bowl. Just before serving, return the wine to a clean saucepan and add the raisins. Heat gently, add the honey, and warm until the wine is coffee hot. Add the fruit slices and serve in heatproof mugs. When tasting, yell "WASSAIL!" to all the beekeepers you are with.

Makes 10 to 20 servings

Roasted Honeyed Plum Tomatoes with Mozzarella and Wild Rocket Salad

The following recipe is enough for 1 serving. Double, triple, or quadruple it as needed.

- ¹/₂ cup (170 g) honey, plus extra for brushing on tomatoes (use a light, mild variety)
- ¹/₂ cup (120 ml) balsamic vinegar
- Extra virgin olive oil
- 1 ¹/₂ plum tomatoes, cut lengthwise
- Rock salt
- 1 teaspoon fresh thyme leaves
- 1 clove garlic, finely chopped
- 1 slice mozzarella cheese, ¹/₂ inch (1.25 cm) thick and large enough to cover half of each tomato slice
- 2 or 3 fresh wild mustard (rocket) leaves

Preheat the oven to 250°F (120°C). Make a balsamic reduction by placing the ¹/₂ cup (170 g) honey and the balsamic vinegar in a small saucepan and heating over medium heat until reduced to the thickness of cooking oil. Brush a baking pan with olive oil and place the tomatoes face down. Brush the tops with honey and sprinkle with rock salt, thyme, and garlic. Drizzle with the balsamic reduction. Roast in the oven for 10 to 12 minutes, being careful to keep the pan in the center of the oven so all sides cook at the same rate. Do not overcook. Remove when hot, but still firm.

To serve, place the tomato halves in the center of a salad plate. Place the cheese on top of the tomatoes. Roll the wild mustard leaves in olive oil and cradle into a ball, passing from one hand to the other. Gently place the leaves on top of the cheese slice. Drizzle the plate with the balsamic reduction by going in and out of the center with one hand, while rotating the plate with the other hand.

Makes 1 serving

Honeyed Pork

ROAST

- **4 teaspoons (20 ml) vegetable oil**
- **1 carrot, roughly cut**
- **1 onion, roughly cut**
- **5-pound (2,275 g) pork loin roast**
- **Salt**
- **1 cup (340 g) honey (use a medium-flavoured variety, such as a midsummer wildflower, or even a goldenrod)**
- **2 tablespoons (30 ml) olive oil**

BASTING

- **1 cup (225 g) butter**
- **1 cup (340 g) honey (same variety as above)**
- **Chives, parsley, rosemary, thyme, for garnish**

For the roast: Preheat the oven to 400°F (200°C). Prepare the roasting pan by pouring in the vegetable oil and the carrot and onion. Score the skin of the roast in a crisscross design with the tip of a very sharp knife. Cut in about ¼ to ½ inch (6 mm to 1.3 cm). Rub salt to taste into the cuts, then rub the honey all over the skin. Warm the olive oil in a large skillet, add the roast, and cook, searing the roast on all sides. When seared, place the roast on top of the vegetables in the pan and place in the middle of the oven. Immediately turn down the heat to 300°F (150°C) and bake for 50 minutes, or 10 minutes per pound (455 g) (15 minutes per pound if you like your pork well, well done).

For the basting mixture: In a small saucepan, combine the butter and honey and set on the stove; the heat from the oven will cause it to melt. Baste the roast about every 10 minutes. Cover the roast with foil for the last 15 minutes, so the roast doesn't get too browned. Remove from the oven and let rest for at least 30 minutes before serving. Garnish wth the herbs.

Makes 10 servings

Mint, Pea, and Basil Puree

Spearmint or peppermint works equally well in this recipe, or for a truly memorable dish, use applemint or chocolate mint leaves. Stick to traditional basil, though, rather than an exotic variety. While there is no honey added, this is a perfect accent for a honey-laden meal.

- **1 sprig mint leaves, stemmed**
- **1 shallot, peeled and finely chopped**
- **1 pound (455 g) fresh peas**
- **5 basil leaves**
- **Salt and freshly ground black pepper**
- **1 cup (235 ml) chicken stock**

In a medium saucepan over medium heat, place the mint leaves, shallot, peas, and basil. Cook until tender. When the shallot is cooked, place the entire mixture in a blender. Add salt and pepper to taste. While blending, slowly pour in the stock. Blend until smooth. Return to the saucepan and warm until ready to serve.

Makes 4 to 5 servings

Earl Grey and Honey Sorbet

- 2 1/2 cups (570 ml) water
- 5 tablespoons (105 g) honey (use the lightest, mildest, flavour you can find)
- 6 tablespoons (188 ml) Earl Grey tea
- Lemon juice from 1/2 lemon
- Mint, for garnish (try spearmint or applemint)

In a medium saucepan, combine the water and honey and bring to a boil. Add the tea and lemon juice and again bring to a boil, reduce the heat, and let simmer for a minute. Filter the mixture through a strainer and taste. Add more lemon juice, if needed, to taste. When just right, pour the mixture into a shallow, flat container with a large surface area on the bottom and place in your freezer. Watch, and when the mixture just begins to freeze (usually 1 to 3 hours, depending on your freezer and the size of your container), remove from the freezer, pour into a blender, and blend on high for 30 seconds, no more. Pour back into the original container and refreeze. When the mixture has the texture of crystallized honey, it is ready to serve. Scoop out using an ice cream or melon scoop, place in a frosted cocktail glass, and garnish with a sprig of mint.

Makes 1 to 2 servings

Roasted Banana

For the best flavour and texture, choose a ripe, but not overly ripe, banana for this recipe.

- 1 banana, unpeeled
- Honey (use a light, mild, sweet variety)
- Mead

Preheat the grill or the oven to 200°F (95°C). Remove one section of the banana's peel and carefully remove the banana. Drizzle a couple of spoonfuls of honey inside the peel and replace the banana. Cut, if needed, to refit, but leave whole if possible. Brush the banana with mead, then replace the missing section of peel. Wrap the banana in foil that has been greased on the inside with a touch of oil and cook on the grill or in the oven for a few minutes, until the banana (inside) is soft, but not mushy.

Makes 1 serving

Glossary

Alarm pheromone: A volatile compound produced by worker honey bees in response to a situation in their environment they find threatening. Alarm pheromone alerts other guard bees in a colony of a potential threat and signals them to be ready to defend home and hearth.

Aldehydes: More than 200 organic compounds found in honey that contribute to its volatile aroma and flavour. Aldehydes are driven off when honey is heated, diminishing its flavour and aroma.

Amino acids: The building blocks of proteins. Amino acids are found in minute quantities in floral nectar and are believed to be an integral component of a plant's physiological system that is contributed to nectar. They are abundant in pollen.

Anthers: The structures on a flower that contain pollen. When mature, anthers open and release pollen in the plant's reproductive process. Anthers open, or dehisce, when environmental conditions favour dispersal. The anther is part of the stamen.

Artisan honey: Artisan honey may be a single varietal honey; that is, it may be derived almost completely from a single plant source, such as from a citrus tree. Or an artisan honey may be a blend of honeys collected from several plant sources, such as from several wildflowers that bloom midsummer. Artisan honey is a unique honey, distinguished by flavour, colour, seasonal production cycle, or other characteristics that make it truly identifiable. For instance, honey collected very early in the spring will most likely be a blend collected from a very few early-blooming tree flowers that generally bloom at about the same time every year. The ratios of these different honeys will differ each year, thus making each year's crop similar, but never the same.

Band heaters: A wide metal band that encircles a pail or drum containing honey. The band has heating elements contained within; when the elements are engaged, the band warms the honey inside the pail or drum. The heat is generally very hot near the band and cooler when more distant, thus overheating the honey closest to the band.

Bee-barren: A location with no plants that provide food for honey bees.

Bee blower: A machine, powered by gasoline or electricity, that propels forced air out of a small nozzle and is used to blow bees out of honey supers during harvest. Bees are not harmed by the forced air.

Bee bread: Pollen mixed with naturally occurring yeasts and honey that ferments and forms an attractive honey bee food. It is stored in combs in the hive and is fed to developing worker larvae.

Bee brush: A soft-bristled brush used to move bees off combs or away from entrances or other places where bees are not needed. Not recommended to move bees off honey combs when harvesting.

Bee-Go: A foul-smelling chemical that serves as a repellent to bees and is used to move bees out of honey supers to be harvested.

Bee-Quick: A chemical that has a pleasant almond scent but it is repellent to bees. It's used to move bees out of honey supers to be harvested.

Bee space: The space—approximately $^3/_8$" (1 cm)— that exists between parts of a hive, such as the frame tops and bottoms, sides, and the like. Smaller spaces will be filled with propolis; larger spaces will be filled with comb. When harvesting, break these seals the day before removing supers to avoid drips and running honey.

Blending: When two or more batches of honey are combined to improve the flavour, colour, or moisture content of the final batch.

Bloom dates: When any particular honey plant is predicted to bloom, using a calendar method of prediction.

Brand melter: A cappings wax-handling machine that melts the beeswax and heats the honey from cappings.

Brood: Any stage of immature honey bee still in the cells.

Brood pheromone: That pheromone produced and given off by all stages of brood that conveys their physical needs to house bees who are feeding the larvae or are capping the cells of pupae. Brood pheromone sends the message to foragers to collect food.

Capped: When nectar has been reduced to less than 18 percent moisture, bees will cover the cell with fresh beeswax, called a capping.

Capping: The beeswax covering over a honey comb cell containing honey.

Cappings scratcher: A many-tined forklike tool used to scratch off the cappings covering honey on honey comb.

Carbohydrates: An organic compound food composed of carbon, hydrogen, and oxygen.

Carrying capacity: The number of bees a particular flower patch can provide pollen and/or nectar for, in total amount produced over time, in amount produced in a specific period, or in room available for foragers.

Chunk honey: A type of honey pack containing one or more pieces of comb honey covered with liquid honey in the same container.

Cold knife: An offset, large-bladed knife that is sharpened on both sides and used to slice cappings off the tops of honey comb cells containing honey. It does not have a heating element as part of the unit. See Hot Knife.

Comb honey: Honey in the comb that has not been extracted. Typically, comb honey is sold intact; that is, it includes beeswax comb, honey, and cappings.

Creamed honey: Finely crystallized honey used as a spread. Crystallization is controlled by the producer using a finely crystallized "seed" and temperature. The technique to accomplish this was patented by E. J. Dyce from Cornell University and is called the Dyce method.

Crystallized honey: Honey that has crystallized either naturally or by the Dyce method.

Curing honey: A process in which bees add appropriate enzymes and use dehydration to convert floral nectar to honey.

Cut-comb honey: Removing a portion of a frame of honey and packaging it as is, including the honey comb, beeswax, and honey still sealed in the cells.

Deeps: The largest super used in a U.S. beehive, measuring $9^9/_{16}$" (24.34 cm) tall. Commonly used as a home for the brood, it often is called a hive body. Commercial beekeepers tend to use deeps for all their boxes to standardize their operations.

Dehydration: The process of reducing floral nectar from its original moisture content down to about 17.5 percent or so, when it becomes honey.

Drift: When honey bees return to their hives in an apiary from foraging trips, they sometimes become confused as to which hive is theirs and return to the wrong hive. If they are allowed in by busy guards, they will most likely take up residence. They have drifted to a new home. Bees tend to drift to end colonies in long, straight rows.

Dry capping: This type of capping occurs when a beeswax capping is placed over a honey comb cell containing honey and there is a tiny air space between the bottom of the wax covering and the surface of the honey. The wax remains white and appears dry.

Enzymes: Catalysts produced by both plants and animals, including the honey bee, that are essential to the chemical reactions in metabolic processes. See Intervase.

Escape board: A device used when harvesting honey. One-way exits allow honey bees to leave a honey super to go to the brood nest, but don't allow them to return to the honey super.

Extra floral nectar: Nectar produced in extra floral nectaries, usually located on leaves or the axils of leaves. Cotton is the classic example of a plant having extra floral nectaries.

Fermentation: When the water content of honey is above 18 percent, the extra water and the sugars from the honey itself allow naturally occurring yeasts in honey to grow, producing, in turn, vinegar, then alcohol. Both render the honey useless.

Filtered: When honey is processed, usually in a commercial packing operation, all particulate matter needs to be removed—pollen grains, dust, wax particles, and the like—everything that glucose crystals can begin to grow on, starting crystallization. To remove even the tiniest particles, honey is heated to 120° to 150°F (49° to 66°C) in a flash heater and pumped at about 5 psi through a series of very fine filters, then flash cooled.

Flail uncapper: Any uncapping machine that uses short chains attached to a spinning axle to strike the surface of a honey comb containing honey to remove the cappings. Horizontal and vertical models are available.

Flash heater: Usually a tank filled with hot water or hot oil through which honey flows in pipes. The surrounding hot material heats the honey as it passes through the tank.

Flower fidelity: When honey bees are on a foraging trip, they visit only one type of flower, thus ensuring pollination. For instance, they will visit only apple blossoms, while ignoring the dandelions on the orchard floor that are simultaneously blooming.

Flower patch: Any group of flowers that are the same. It may be a field of dandelions, a locust tree, or a whole field of honeysuckle bushes.

Foragers: Honey bees that leave the hive to harvest nectar, pollen, propolis, or water for their colony.

Fructose: One of the two main sugars of honey, the other being glucose. These are both six-carbon sugars, derived from sucrose, a twelve-carbon sugar when invertase is added by the bees.

Fume board: A top cover for a beehive that has an absorbent pad on the underside, used with a liquid repellent that causes the bees to move from the honey super below. See Bee-Go and Bee-Quick.

Gluconic acid: The principal acid in honey.

Glucose: One of the two main sugars in honey. It is a six-carbon sugar, derived from sucrose, a twelve-carbon sugar. The other main sugar in honey is fructose.

Growing degree days: The figure derived from adding the low temperature of the day to the high temperature of the day and dividing by two to obtain the average daily temperature, then subtracting the low temperature of the day from that average. The resulting figure is the number of growing degree days that particular day produced. Used to predict with great accuracy the bloom date of most crop and honey plants.

Guards: Honey bees that have matured to the point of being able to fly well; their sting glands have matured also.

Herbicides: Chemicals that are applied to plants by farmers or anyone wishing to kill the plants.

High-fructose corn syrup (HFCS): A sugar syrup, made from corn, that contains glucose and fructose. It's used for feeding honey bees and as an additive in nearly everything else we as humans eat. Available in high and low concentrations.

Honeydew: A sugary liquid excreted by aphids and scale insects (usually those in the order Homoptera). Bees will collect it and ripen it as they would floral nectar and turn it into a type of honey.

Honey flow: The term given to the time when a large number of nectar-producing plants are in bloom.

Honey stomach: An enlarged area at the posterior end of a honey bee's esophagus but lying in the front part of the abdomen, capable of expanding when filled with nectar or water. It has no digestive properties and serves solely as a holding space for transporting nectar or water.

Hot knife: An offset, large-bladed knife that is sharpened on both sides and used to slice cappings off the tops of honey comb cells containing honey. It has a heating element as part of the unit to warm the blade. Some have a thermostat set in the blade, others in the wire; still others have a graduated temperature gauge attached to the unit for best control. See Cold Knife.

Hot room: An area where honey supers are stored, usually at or very near the uncapping facility, after they have been harvested and before they are processed. Ideally, it has an ambient temperature between 80° and 100°F (27° and 38°C). The honey is kept warm, or warmed, to facilitate extracting.

House bees: Adult honey bees that are not yet fully developed. Generally, they are not yet able to fly or sting, and are engrossed in such activities as cleaning cells, drying honey, feeding larvae, and caring for the queen.

Hypopharyngeal gland: Located in the heads of worker bees, these glands produce royal jelly and brood food.

Invertase: The enzyme added to nectar by honey bees that aids in the change of sucrose as a twelve-carbon sugar to glucose and fructose, six-carbon sugars.

Larva: The second stage of honey bee metamorphosis, which is complete metamorphosis. It is the grub or worm stage, preceded by the egg stage, followed by the pupa, or cocoon stage, followed by the adult stage.

Medium supers: Boxes used in beekeeping primarily for honey storage. Medium supers are added above the hive bodies for storage space for the bees to put their honey in. They are 6 5/8" (17 cm) tall.

Monoculture: An area that is vegetatively dominated by a single plant species. A corn field without weeds or other plants along fencerows would be a monoculture.

Nasonov pheromone: This pheromone attracts workers to food, a nest site, or a new home. Used in organising swarm movement. The Nasonov gland is found on the basal part of tergum seven of the worker bee.

Nectar: A sugary liquid secreted by special glands called nectaries located chiefly in flowers and also by extra floral nectaries. Though it is comprised of water and primarily sucrose sugar, other sugars, enzymes, proteins, and materials are also found there in small amounts.

Nectaries: Organs of a plant made up of specialized tissues that secrete nectar.

Parallel radial extractor: A machine used to remove honey from honey comb. It is organized much like a bicycle tire, with the honey frames in the spoke positions. It operates in a vertical plane such that the frames rotate exactly like a wheel, rather than a potter's wheel, which is a horizontal radial extractor.

Pfund grader: A machine that measures honey colour by how much light can pass through a sample of known thickness. Measured in millimeters, ranging from 0 to 114, with the largest numbers representing the darkest colours.

Pheromone: A substance secreted by insects that when sensed or ingested by other individuals of the same species causes them to respond with a definite behavior or developmental process.

Pollination: The transfer of pollen from the anther to the stigma of a flower.

Proboscis: The tongue or combined maxillae and labium of a bee. The mouthparts of the bee that form the sucking tube and tongue.

Propolis: Sap or resinous material collected from the buds or wounds of plants by bees and used to strengthen wax comb, seal cracks, reduce entrances, and smooth rough spots in the hive.

Queen excluder: Any device having openings permitting the passage of worker bees but excluding the passage of drones and queen bees. Prevents the queen from entering honey supers.

Queen pheromone: A substance, composed of many individual chemicals, produced by a queen that the attendant worker bees collect and pass to the rest of the colony. If the queen's supply of the secretion is not adequate or ceases entirely, or the ratio of the individual substances changes dramatically, the colony will be motivated to produce laying workers, or swarm cells, or supersede the existing queen.

Races of bees: A commonly used term for honey bee subspecies, the most common being Italian, Carniolan, and African.

Radial extractor: A machine designed to remove honey from a honey comb. Constructed to resemble a bicycle wheel, with the top bars facing the outside edge in an upright position.

Recruit (noun): A honey bee that has been convinced that she should follow a forager's instructions to where a profitable flower patch is.

Recruit (verb): To enthusiastically dance and offer samples of nectar or pollen that come from a profitable flower patch to other foragers for taste and smell. The purpose of recruiting is to convince the other foragers to go to that patch.

Refractometer: An instrument that measures the water content of any number of liquid samples by measuring the amount of solids in the sample, but offering the remaining material as the measured faction, as a percent.

Ripe honey: Honey that has been reduced to less than 18 percent moisture content.

Ripening: The process honey bees go through to reduce nectar from sucrose at a high moisture content, to honey with a low moisture content. This is achieved by adding enzymes that fractionate the sucrose molecule and reducing the water content so it is below 18 percent.

Ross Rounds: These plastic frames are constructed such that they produce comb honey in their own container. The beekeeper inserts thin beeswax foundation between the frame halves and lets the bees build comb, fill the comb with honey, and cap the honey when done. The beekeeper then removes the self-contained round sections from the frames and places a cap on each side of the section and a label around the circumference. The comb honey is then ready to give or sell.

Royal jelly: The food produced by house bees to feed queen larvae. It is a milky white, thick liquid secreted from the hypopharyngeal glands of nurse bees and is high in sugar and protein.

Scout bees: Scout bees seek out flower patches and stop and sample the flowers' nectar and pollen to measure its profitability in terms of sugar content and amount (for nectar) and protein richness (for pollen). They then return to the hive and recruit other foragers to visit the patch. Their recruiting vigor is dictated by the value of the patch.

Settling: Because honey is a dense liquid, particles suspended in it will eventually rise to the top, but it takes some time. Air bubbles, pollen, dust, and other material will all rise and float on the surface of a container of honey. This is called settling.

Shallow supers: These are the smallest supers that hold traditional frames and are used to collect surplus honey. They measure 5³/₄" (14.6 cm) deep.

Small hive beetle: This beehive pest was accidentally introduced into the United States from its native South Africa. *Aethina tumida* invades beehives in the field and consumes eggs, larvae, wax, honey, and pollen, defecating in the hive as it eats. It is very destructive in hives, and especially in stored honey combs in honey houses.

Strained: Honey is strained by running it through fine-mesh screens or cloths using gravity. It is not previously flash heated. Strained honey generally needs to be settled.

Sucrose: Sucrose is essentially table sugar. It is by far the predominant sugar found in floral nectar.

Supersaturated: A supersaturated solution is one where more material is dissolved in the liquid than would normally be; this is a result of extreme temperature, pressure, or other factors. When the solution returns to normal, some of the dissolved material will come out of solution.

Surplus honey: This is the honey that beekeepers do not harvest and that is used by the bees for food, generally for overwintering.

Tangential extractor: A tangential extractor spins frames such that the comb, which sits in a basket inside the extractor barrel, faces the inside of the extractor. The extractor basket is spun such that the honey on one side of a frame is thrown out. Then the frame is turned 180 degrees so the other side of the frame faces the inside of the extractor and it is spun again, throwing the rest of the honey out of the frame.

Uncapped honey: Nectar that has not yet reached that magical 18 percent or less moisture level will not be capped, even though it looks like honey, or actual honey that has fallen below that amount but has not yet been capped.

Uncapping fork: See Cappings Scratcher.

Uncapping knives: Specially designed tools featuring offset handles and broad, sharp blades that are not flexible. These knives are used to remove the beeswax cappings from capped honey.

Uncapping machines: Machines that remove the beeswax cappings from capped honey frames using heated, oscillating blades; short, spinning chains that strike the cappings as a frame passes by, removing the cappings as they are struck; punches that poke holes in the cappings; or revolving blades that cut away the cappings as the frames pass between, under, or over these devices.

Varietal honey: Honey that is harvested that has been collected by the bees from a single floral source, or mostly a single source. Some resources claim the honey must be at least 51 percent pure, while others push for far more purity. Though it's difficult to obtain 100 percent pure varietal honey, only something in the neighborhood of at least 80 percent or more should be considered varietal honey, in the opinion of the author.

Varroa mites: *Varroa destructor*, the varroa mite, is by far the worst enemy of honey bees, and thus beekeepers, on the planet. It is an external mite that parasitizes honey bee larvae and adults, aids in the transmission of virus diseases, kills larvae and adults when infestations become high, and overall has killed more honey bees than any other pest or predator, ever.

Ventriculus: The mid-gut, or stomach, of a honey bee, located in the abdomen between the honey stomach (proventriculus) and the hindgut.

Volatile ketones: Aromatic compounds found in very small quantities in honey that contribute to the flavours and aromas of various honey varieties. They are extremely volatile and are easily driven off when honey is heated.

Warming box: Any device in which containers of honey are placed that provides a warming environment that hastens melting crystallized honey.

Wax moth: A moth whose larvae destroy honey combs by boring through the wax in their search for food, leaving trails of webbing and collections of frass, shed skins, and destruction. There are two species, the greater and the lesser wax moth. Both are destructive in hives and particularly in stored honey and brood supers.

Wax spinner: A machine that separates beeswax residue from honey using centripetal force, much like a salad spinner. The honey is driven off when the inside cylinder spins, leaving the wax inside and the honey to drain away.

Wet cappings: When honey has ripened, or reached 18 percent moisture or less, honey bees place a beeswax cover over the cell the honey is in to keep it clean. Some bees place this covering so that there is a small air space between the top of the honey and the bottom of the covering, or cap. The wax then appears to be white. If the beeswax covering lies directly on the surface of the honey, the wax will appear to be wet, thus wet cappings.

Resources

BOOKS

A Honey Cook Book: Recipes from the Home of the Honey Bee. A. I. Root Co.

Aston, David, and Sally Bucknall. *Plants and Honey Bees.* F. N. Howes, 2000.

Benjamin, Alison, and Brian McCallum. *Keeping Bees and Making Honey.* David and Charles Publications, 2008.

Blackiston, Howland. *Beekeeping for Dummies.* Hungry Minds Press, 2002.

Buchmann, Stephen L., and Gary Paul Nabhan. *The Forgotten Pollinators.* Island Press, 1997.

Caron, Dewey, and Thomas Webster. *Observation Hives.* A. I. Root Co., 2000.

Conrad, Ross. *Natural Beekeeping.* Chelsea Green Publishing, 2007.

Crane, Eva. *The World History of Beekeeping and Honey Hunting.* Routledge, 1999.

Davis, Celia F. *The Honey Bee Inside Out.* Bee Craft Ltd., 2004.

Delaplane, Keith. *First Lessons in Beekeeping.* Dadant and Sons, 2007.

Flottum, Kim. *Quick and Easy Guide to Beekeeping.* Apple Press, 2005.

Goodman, Lesley. *Form and Function in the Honey Bee.* International Bee Research Association, 2003.

Graham, Joe (editor). *The Hive and the Honey Bee.* 4th ed. Dadant and Sons, 1975.

Harman, Ann, H. Shimanuki, and Kim Flottum (editors). *The ABC and XYZ of Bee Culture.* 41st ed. A. I. Root Co., 2007.

Hooper, Ted, and Mike Taylor. *The Beekeeper's Garden.* Alphabooks, 1988.

Langstroth, L. L. *Langstroth's Hive and the Honey-Bee.* 4th ed. Dover Books, 2004.

Lovell, John H. *Honey Plants of North America.* A. I. Root Co., 1999.

Morse, Roger A., and Kim Flottum. *Honey Bee Pests, Predators, and Diseases.* 3rd ed. A. I. Root Co., 1998.

Munn, Pamela, and Richard Jones (editors). *Honey and Healing.* International Bee Research Association.

Niall, Mani. *Covered in Honey.* Rodale, 2003.

Ramsay, Jane. *Plants for Beekeeping in Canada and North America.* International Bee Research Association.

Seeley, Thomas D. *The Wisdom of the Hive.* Harvard University Press, 1996.

Traynor, Joe. *Honey: The Gourmet Medicine.* Kovak Books, 2002.

University of Wyoming (editors). *Weeds of the West.* Western Society of Weed Science.

Uva, Richard H., Joseph C. Neal, and Joseph M. Ditomaso. *Weeds of the Northeast.* Cornell University Press, 1997.

von Frisch, Karl. *The Dance Language and Orientation of Bees.* Harvard University, 1967.

MAGAZINES

UK and Ireland

An Beachaire
The Irish Beekeeper
www.irishbeekeeping.ie

BBKA News
British Beekeepers' Association
bbka@britishbeekeepers.com

BeeCraft
British Beekeepers' Association
www.bee-craft.com

The Beekeepers Quarterly
www.beedata.com

Beekeeping
www.devonbeekeepers.co.uk

Bees for Development Journal
www.beesfordevelopment.org

The Journal of Apicultural Research
www.IBRA.org.uk

The Scottish Beekeeper
www.scottishbeekeepers.org.uk

Europe

Aricilik Dergisi
Uludag Bee Journal
www.uludagaricilik.org.tr

Beekeeper
The Magazine of Serbian Beekeeping
www.spos.info

Bienen-Zietung
Schweizerische
www.vdrb.ch

Birokteren
Norwegian Beekeeping
www.norbi.no

Bitidningen
Swedish Beekeeping
www.biodlarna.se

Deutsches Bienen Journal
bienenjournal@bauernverlagl.de

Mehilainen
Finnish Beekeeping
www.hunaja.net

Pasieka
www.pasieka.pszczoly.pl

La Sante de L'Abeille
www.sante-de-labeille.com

Tidskrift for Biavl
Danish Beekeeping
www.biavl.dk

Vida Apicola
www.vidaapicola.com

North America

American Bee Journal
Dadant and Sons
www.dadant.com

Bee Culture, The Magazine of American
Beekeeping
A. I. Root Co.
www.BeeCulture.com

Hive Lights
Canadian Honey Council
www.honeycouncil.ca

Australia

The Australasian Beekeeper
www.penders.net.au

Australian Bee Journal
abjeditors@yahoo.com

SUPPLIERS OF BEEKEEPING EQUIPMENT

Bee Equipped
Ashbourne, Derbyshire, UK
44 01335 370567
freespace.virgin.net/bee.equipped/index.htm

Bees-online
Gloucestershire, UK
44 01452 700289
www.bees-online.co.uk

B J Sherriff
Falmouth, Cornwall, UK
44 1872 863304
www.bjsherriff.com

E. H. Thorne
Wragby, UK
www.thorne.co.uk

Fragile Planet Ltd
Oswestry, Shropshire, UK
44 01691 672869
www.fragile-planet.co.uk

Guilfoyle Beekeeping Equipment
Australia
guilfoylewa@tnet.com.au

Kemble Bee Supplies
Hastings, East Sussex, UK
44 1424 870729
www.kemble-bees.com

Medivet
Alberta, Canada
www.medivet.ca

Michael Jay Beekeeping Supplies
Temple Cloud, Somerset, UK
44 0 1761 452344

Modern Beekeeping Limited
Kingsbridge, Devon, UK
44 01548 858747
www.modernbeekeeping.co.uk

National Bee Suppliers
Oakhampton, Devon, UK
44 (0) 1837 54084
www.beekeeping.co.uk

Paynes Southdown Bee Farms Ltd
Hassocks, West Sussex, UK
44 01273 843388
www.paynesbeefarm.co.uk/store/

Penders Beekeeping Supplies
Australia
(02) 4956 6166

Swienty
Sweden
www.swienty.com

The Honey Pot
Markeaton Lane, Derby, UK
44 01332 203 893
www.localhoney.co.uk

Thomas Apiculture
France
www.thomas-apiculture.com

Index

Photographer Credits

All photographs by Kim Flottum with the exception of the following:

Courtesy of Brushy Mountain Bee Farm, 91; 98 (top); 104 (middle)

Courtesy of Cowan Manufacturing Co., Inc., 92 (top)

Courtesy of Dadant & Sons, Inc., 93; 98 (bottom); 99 (bottom); 106; 112 (bottom); 113

www.fotolia.com, 48

www.istockphoto.com, 6 (left & right); 8; 15 (middle, top); 17 (top); 19 (top, right); 25; 43 (top); 49; 51; 61; 75; 79; 120

Courtesy of Walter J. Kelley Co., 95

Courtesy of Mann Lake Ltd., 86; 88; 89; 96

Courtesy of Maxant, 99 (top); 112 (top)

Courtesy of Park, 55; 57

Glenn Scott Photography, 2; 121; 122–167

Kathy Summers, 168

Courtesy of the United States Department of Agriculture, 27; 54 (top)

About the Author

Kim Flottum is the editor of *Bee Culture* magazine, published by the A. I. Root Company in Medina, Ohio, where he has been for more than twenty years. His book *Complete and Easy Guide to Beekeeping* was published by Apple Press in 2005.

He edited (with Dr. Shiminuki and Ann Harman) the popular 41st edition of the *ABC and XYZ of Bee Culture*, the bible of U.S. beekeeping published by the A. I. Root Company.

Kim blogs for www.thedailygreen.com on the beekeeping world, plus he is a regular columnist for the U.K. beekeeping magazine *The Beekeeper's Quarterly*. He also writes articles on the beekeeping life for many agricultural and gardening journals.

Kim and his wife Kathy keep a few hives in the backyard, while a garden, two cats, hundreds of exotic plants, and extensive travel for his job and volunteer efforts fill much of his time.

Kim's best advice to beekeepers? Keep your smoker lit and your hive tool sharp, because next year *will* be better.

Acknowledgments

There are many people who contribute to a project such as this simply because it is the right thing to do for those who will someday read, learn from, and enjoy the final product.

Thank you to the friends and associates who contributed their photos of the equipment, including Brushy Mountain, Mann Lake Ltd., Dadant and Sons, Maxant, Cowen Manufacturing, Walter T. Kelley Company, Dakota Gunness Uncappers, ooA. I. Root Company, and the others mentioned in the photographer credits. This work could have been done without their help, but it would have been grossly incomplete.

Individuals, too, have contributed to this work, some knowingly, others anonymously. These include Buzz Riopelle, Roy Hendrickson, Bob Smith, and scores of beekeepers who let me wander into their lives, ask questions, and take photos. Thank you.

Michael Young's recipe contributions must be acknowledged. He is a master chef and a pretty good beekeeper combined. I'm a kitchen jockey who loves honey, simply experimenting and enjoying my trials and errors. Michael, photographer Glenn Scott, photo stylist Catrine Kelty, and the design department at Quarry Books are the true artists here. In the process, they made the recipes you see here simply sing.